T0305657

Statistical Methods Using SPSS

Statistical Methods Using SPSS provides a practical approach for better understanding of the advanced statistical concepts that are applied in business, economics, epidemiology, public health, agriculture and other areas of data analytics. Advanced statistical methods or advanced statistical techniques for analyzing data arise because of the complex nature of data sets that cannot be analyzed using the basic or the usual and common analytical techniques. This book describes more advanced statistical methods, offering a modern approach by introducing the advanced statistical concepts, before showing the application of these concepts in real-world examples with the application of SPSS statistical software.

This book is useful in explaining advanced statistical analysis techniques to postgraduate students, doctoral students and researchers. It is also a useful reference for students and researchers who require further guidance in advanced data analysis and is designed for those with basic statistical knowledge. Exercises are also included at the end of each chapter to aid in the understanding of the statistical analysis techniques explained in the book.

Key features:

- there are many topics on advanced statistical techniques,
- a provision of theoretical statistical concepts,
- there is a step-by-step guide for the different statistical analysis techniques being done using SPSS,
- there are variety of data set examples to help explain the different statistical concepts, and
- there is a practical applications of the statistical concepts in SPSS.

Gabriel Otieno Okello is a member of the statistics and data science faculty at the School of Science and Technology, United States International University, Nairobi, Kenya. Currently, he is an assistant professor of statistics and data science. He is also a senior statistics and data science consultant.

Statistical Methods Using SPSS

Gabriel Otieno Okello

CRC Press is an imprint of the
Taylor & Francis Group, an **informa** business
A CHAPMAN & HALL BOOK

Designed cover image: © Gabriel Otieno Okello

First edition published 2025
by CRC Press
2385 NW Executive Center Drive, Suite 320, Boca Raton FL 33431

and by CRC Press
4 Park Square, Milton Park, Abingdon, Oxon, OX14 4RN

CRC Press is an imprint of Taylor & Francis Group, LLC

Library of Congress Cataloging-in-Publication Data
Names: Otieno Okello, Gabriel, author.
Title: Statistical methods using SPSS / Gabriel Otieno Okello.
Description: First edition. | Boca Raton, FL : CRC Press, 2024. |
Includes bibliographical references and index.
Identifiers: LCCN 2023054022 (print) | LCCN 2023054023 (ebook) |
ISBN 9781032479361 (hbk) | ISBN 9781032473093 (pbk) | ISBN 9781003386636 (ebk)
Subjects: LCSH: SPSS (Computer file) | Social sciences--Statistical methods--Data processing. | Statistics--Data processing.
Classification: LCC HA32 .O85 2024 (print) | LCC HA32 (ebook) |
DDC 519.5--dc23/eng/20240314
LC record available at https://lccn.loc.gov/2023054022
LC ebook record available at https://lccn.loc.gov/2023054023

ISBN: 978-1-032-47936-1 (hbk)
ISBN: 978-1-032-47309-3 (pbk)
ISBN: 978-1-003-38663-6 (ebk)

DOI: 10.1201/9781003386636

Typeset in Minion
by KnowledgeWorks Global Ltd.

Contents

CHAPTER 1

Introduction to Advanced Statistical Techniques

1.1 INTRODUCTION

Advanced statistical methods or advanced statistical techniques for analyzing data arise because of the complications in analytical techniques for simple data sets.

Usually data can be categorized as numeric or non-numeric data. Numeric data are data that are quantitative in nature, while non-numeric data are data sets that are either categorical or qualitative in nature. Statistical analysis usually begins with simple to advanced analyses or analyses that are based on some specified assumptions such as normal distribution to analyses that go beyond the specified distributional assumptions. Analysis also begins with univariate, then bivariate and multivariate analyses.

Whenever we have situations that are beyond numeric data, or when the specified assumptions for the given analyses are violated, then we may need advanced statistical techniques to be able to analyze that data.

We can have situations where data set is only categorical, and we would like to predict some categorical outcomes based on a combination of both categorical and numeric explanatory variables.

We may also have situations where we are dealing with multivariate variables, time to event data (survival data) or even longitudinal or panel

DOI: 10.1201/9781003386636-1

or repeated measures or spatial data. This type of data requires specialized analytical techniques.

We may also have situations where we have missing data, and we would like to have statistical techniques that handle missing data. We may also have georeferenced/spatial data.

There are also situations where we have the assumptions for a given analytical technique not holding nor being met, and this will require specialized techniques for analyzing those data sets.

In this book, Statistical Methods Using SPSS, we present specialized techniques that can be used to analyze data that are in the situations stated above.

Statistical Methods Using SPSS book will provide a practical approach for better understanding of the advanced statistical concepts that are applied in business, economics, epidemiology, public health, agriculture and other areas of data analytics.

In this book, we will illustrate the step-by-step procedures for performing advanced statistical analysis. It will be useful to postgraduate students, doctoral students and researchers. This book can be used as a very useful reference for doctoral, postgraduate and other high-level courses for undergraduate and other researchers who require guide in data analysis. The book has SPSS screenshots that guide and offer explanations to the users.

1.2 PRACTICE EXERCISE

1. Describe the need for advanced statistical techniques for data analysis.
2. How will you categorize the various advanced statistical data analysis techniques?

CHAPTER 2

Multivariate Analysis Using SPSS

2.1 INTRODUCTION

Multivariate data arise when researchers measure several variables on each "unit" in their sample. Multivariate data are data that have several variables with several observations.

The goals of multivariate analysis include description of the p-dimensional distribution (i.e. multivariate means, variances, covariances and multivariate probability distributions), data reduction (i.e. reducing the number of variables without losing significant information), data grouping or sorting (i.e. clustering and classification), investigating relationships/dependence among variables of interest and statistical inference (i.e. confidence regions, multivariate regression, hypothesis testing).

Multivariate analysis techniques can be generally classified into two broad categories: dependence methods and interdependence methods. For the dependence methods, one type for data has both dependent and independent variables, while for the interdependence methods, the type for data has several variables without dependency relationship.

Dependence methods include multiple regression analysis, multiple discriminant analysis, multivariate analysis of variance and canonical analysis, whereas the interdependence methods include factor analysis, cluster analysis, multidimensional scaling or MDS (both metric and nonmetric) and the latent structure analysis.

DOI: 10.1201/9781003386636-2

Most multivariate data sets have a common form and consist of a data matrix, the rows of which contain the units in the sample/observations and the columns of which refer to the variables measured on each unit. Symbolically a set of multivariate data can be represented by the matrix, X, given by:

$$X = \begin{bmatrix} x_{11} & \cdots & x_{1q} \\ \vdots & \ddots & \vdots \\ x_{n1} & \cdots & x_{nq} \end{bmatrix}.$$

where:

n is the number of units in the sample.

q is the number of variables measured on each unit.

x_{ij} denotes the value of the jth variable for the ith unit.

2.2 VISUALIZING MULTIVARIATE DATA

An important step in the analysis of any data set is exploratory data analysis (EDA), which includes the graphical display of data.

Graphical display of data is important because:

- It may suggest a plausible model for the data.
- It assess the validity of model assumptions.
- It helps detect outliers.
- It may suggest plausible normalizing transformations.

With multivariate data we use the following techniques to visualize multivariate data.

2.2.1 A Matrix Scatterplot

One common way of plotting multivariate data is to make a "matrix scatterplot", showing each pair of variables plotted against each other.

A matrix scatterplot is used in multivariate cases to:

- Look at all the relationships between pairs of variables in one group of plots.
- Describe pairwise relationships among three or more variables simultaneously.

A "matrix scatterplot" shows how each pair of variables is plotted against each other.

To generate a scatterplot matrix in SPSS, we go to: Graphs > Chart Builder OK > select Scatter/Dot under the Gallery to choose from: > select Scatter/Dot under the Gallery to choose from: > select the last scatter plot (Scatterplot Matrix) and drag it to the top window beside the Variable box > select the multivariate variables from the Variables box and drag them to the Scatterplot Matrix then click OK. SPSS Output will generate a scatterplot matrix.

Example 2.1: Consider the following data (Figure 2.1). Generate a scatterplot matrix.

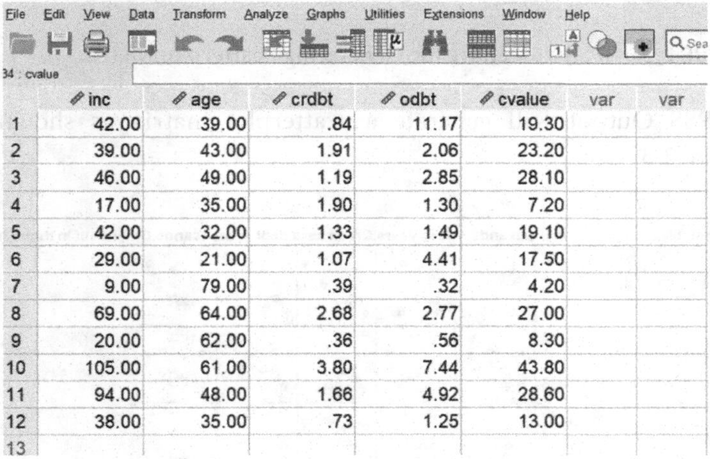

	inc	age	crdbt	odbt	cvalue	var	var
1	42.00	39.00	.84	11.17	19.30		
2	39.00	43.00	1.91	2.06	23.20		
3	46.00	49.00	1.19	2.85	28.10		
4	17.00	35.00	1.90	1.30	7.20		
5	42.00	32.00	1.33	1.49	19.10		
6	29.00	21.00	1.07	4.41	17.50		
7	9.00	79.00	.39	.32	4.20		
8	69.00	64.00	2.68	2.77	27.00		
9	20.00	62.00	.36	.56	8.30		
10	105.00	61.00	3.80	7.44	43.80		
11	94.00	48.00	1.66	4.92	28.60		
12	38.00	35.00	.73	1.25	13.00		
13							

FIGURE 2.1 Sample Multivariate Data

To generate Scatter Plot Matrix in SPSS, we go to: Graphs > Chart Builder OK > select Scatter/Dot under the Gallery to choose from: > select the last scatter plot (Scatterplot Matrix) and drag it to the top window beside the Variable box > select the multivariate variables (inc, age, crdbt, odbt, cvalue) (Figure 2.1) from the Variables box and drag them to the Scatterplot Matrix then click OK (Figure 2.2).

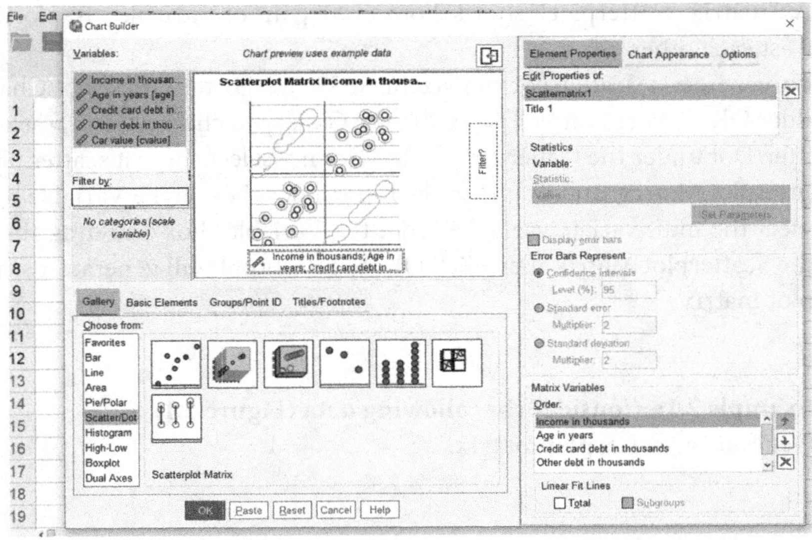

FIGURE 2.2 SPSS Procedure for Generating Scatterplot Matrix

SPSS Output will generate a scatterplot matrix as shown in Figure 2.3.

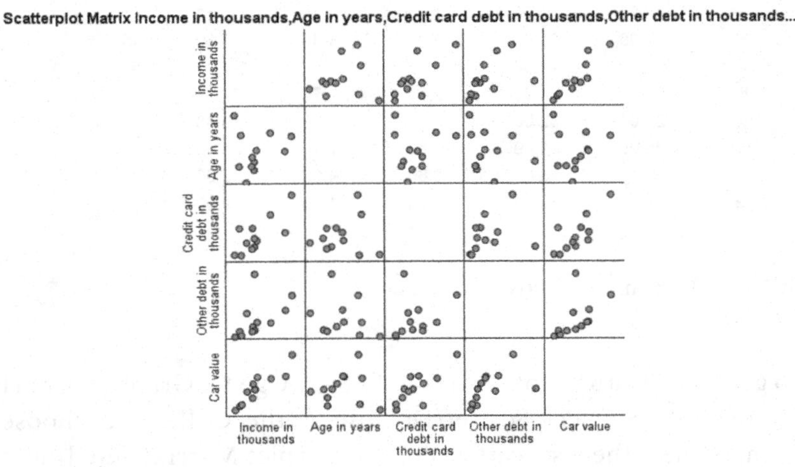

FIGURE 2.3 Scatterplot Matrix

The scatterplot matrix presented in Figure 2.3 shows the strong and linear relationships between the variables.

2.2.2 A Profile Plot

Another type of plot that is useful is a "profile plot", which shows the variation in each of the variables, by plotting the value of each of the variables for each of the samples. Profile plot is a line plot in which each point indicates the estimated marginal mean of a dependent variable (adjusted for any covariates) at one level of a factor.

A profile plot is used in multivariate case to show variation in each of the variables, by plotting the value of each of the variables for each of the samples. Profile plots are generated when performing multivariate analysis of variance (MANOVA).

2.3 SUMMARY STATISTICS FOR MULTIVARIATE DATA

In order to summarize a multivariate data set, we need to produce summaries for each of the variables separately and also to summarize the relationships between the variables. The other thing that you want to do is to calculate summary statistics such as the mean and standard deviation for each of the variables in your multivariate data set.

Three measures for summarizing multivariate data exist: measures of central tendency, measures of dispersion and measures of association.

2.3.1 Measures of Central Tendency

Measures of central tendencies for multivariate data are used to summarize the central values in a data set with multiple variables. The most commonly used measure is the mean vector.

2.3.1.1 Mean Vector

For q variables, the population mean vector is usually represented as $\mu' = [\mu_1, \mu_2, \ldots, \mu_q]$

where $\mu_i = E(x_i)$ is the population mean (or expected value as denoted by the E operator in the above) of the ith variable. An estimate of μ', based on n, q-dimensional observations, is $\bar{x}' = [\bar{x}_1, \bar{x}_2, \ldots, \bar{x}_q]$, where $\bar{x}_i$ is the sample mean of the variable x_i.

The population mean vector, μ, is a collection of the means for each of the population means for each of the different variables.

2.3.2 Measures of Dispersion

Measures of dispersion for multivariate data provides insights into how the values of each variable vary across observations. The most commonly used measure of dispersion for multivariate data is variance-covariance matrix.

2.3.2.1 Variance

A variance measures the degree of spread (dispersion) in a variable's values.

Population variance is the average squared difference between a variable's values and the mean for that variable.

The vector of population variances can be represented by $\sigma' = \left[\sigma_1^2, \sigma_2^2, \ldots, \sigma_q^2\right]$

where

$$\sigma_i^2 = E\left[x_i - \mu_i\right]^2 .$$

An estimate of σ' based on n, q-dimensional observations is

$$s' = \left[s_1^2, s_2^2, \ldots, s_q^2\right].$$

where s_1^2 is the sample variance of x_i.

2.3.3 Measures of Association

2.3.3.1 Covariance

The population covariance is a measure of the association between pairs of variables in a population.

The population covariance of two variables, x_i and x_j, is defined by

$$Cov\left(x_i, x_j\right) = E\left(x_i - \mu_i\right)\left(x_j - \mu_j\right).$$

Recall that we had collected all the population means of the p variables into a mean vector. Likewise, the population variances and covariances can be collected into the **population variance-covariance matrix**. This is also known by the name of **population dispersion** matrix.

The covariance of x_i and x_j is usually denoted by σ_{ij}.

$$\Sigma = \begin{pmatrix} \sigma_{11} & \sigma_{12} & \sigma_{1q} \\ \sigma_{21} & \sigma_{22} & \sigma_{2q} \\ & & \\ \sigma_{q1} & \sigma_{q2} & \sigma_{qq} \end{pmatrix}.$$

This matrix is generally known as the variance–covariance matrix or simply the covariance matrix. The matrix Σ is estimated by the matrix, S, given by

$$S = \frac{1}{n-1}\sum_{i=1}^{n}\left(x_i - \bar{x}\right)\left(x_i - \bar{x}\right)' .$$

where $x_i' = [x_{i1}, x_{i2}, \ldots, x_{iq}]$ is the vector of observations for the ith individual. The diagonal of S contains the variances of each variable.

2.3.3.2 Correlations

The covariance is often difficult to interpret because it depends on the units in which the two variables are measured; consequently, it is often standardized by dividing by the product of the standard deviations of the two variables to give a quantity called the correlation coefficient, ρ_{ij}, where

$$\rho_{ij} = \frac{\sigma_{ij}}{\sqrt{\sigma_{ii}\sigma_{jj}}}.$$

For sample data, the correlation matrix contains the usual estimates of the ρ's, namely Pearson's correlation coefficient, and is generally denoted by R. The matrix may be written in terms of the sample covariance matrix S as follows:

$$R = D^{-\frac{1}{2}} S D^{-\frac{1}{2}}.$$

where $D^{-\frac{1}{2}} = diag(1/s_i)$.

2.3.3.3 Distances

The concept of distance between observations is of considerable importance for some multivariate techniques. The most common measure used in Euclidean distance, for two rows, namely row i and row j, of the multivariate data matrix, X, is defined as:

$$d_{ij} = \left[\sum_{k=1}^{q} (x_{ik} - x_{jk})^2\right]^{\frac{1}{2}}.$$

To generate multivariate statistics in SPSS, we go to: Analyze > Correlate > Bivariate > select and move variables of interest to Variables box > click Options and select Cross-product deviations and covariances then click Continue then OK. SPSS Output will display the summary statistics for multivariate data.

Example 2.2: Consider the data in Figure 2.1. Estimate the multivariate statistics.

We go to: Analyze > Correlate > Bivariate > select and move the five variables (inc, age, crdbt, odbt, cvalue) to Variables box > click

Options and select Cross-product deviations and covariances then click Continue then OK (Figure 2.4).

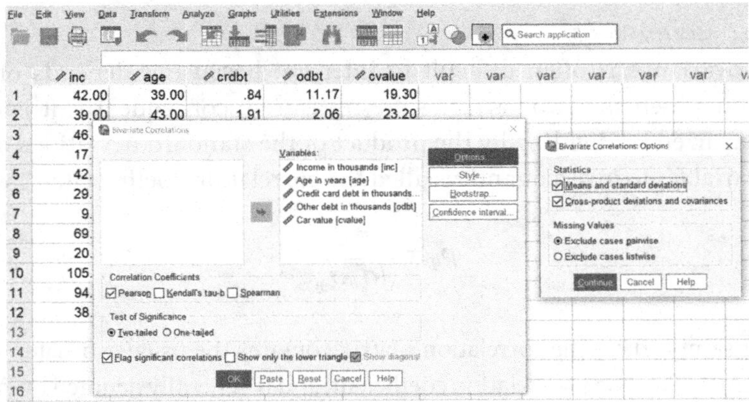

FIGURE 2.4 SPSS Procedure for Generating Multivariate Summary Statistics

SPSS Output will display the summary statistics for multivariate data as shown in Figure 2.5.

Descriptive Statistics

	Mean	Std. Deviation	N
Income in thousands	45.8333	29.59371	12
Age in years	47.3333	16.53280	12
Credit card debt in thousands	1.4879	.99572	12
Other debt in thousands	3.3781	3.20232	12
Car value	19.9417	11.16866	12

Correlations

		Income in thousands	Age in years	Credit card debt in thousands	Other debt in thousands	Car value
Income in thousands	Pearson Correlation	1	.130	.751**	.510	.913**
	Sig. (2-tailed)		.688	.005	.090	<.001
	Sum of Squares and Cross-products	9633.667	697.667	243.496	531.601	3320.383
	Covariance	875.788	63.424	22.136	48.327	301.853
	N	12	12	12	12	12
Age in years	Pearson Correlation	.130	1	.116	-.169	.052
	Sig. (2-tailed)	.688		.719	.600	.873
	Sum of Squares and Cross-products	697.667	3006.667	21.059	-98.414	105.033
	Covariance	63.424	273.333	1.914	-8.947	9.548
	N	12	12	12	12	12
Credit card debt in thousands	Pearson Correlation	.751**	.116	1	.304	.785**
	Sig. (2-tailed)	.005	.719		.337	.002
	Sum of Squares and Cross-products	243.496	21.059	10.906	10.649	96.076
	Covariance	22.136	1.914	.991	.968	8.734
	N	12	12	12	12	12
Other debt in thousands	Pearson Correlation	.510	-.169	.304	1	.543
	Sig. (2-tailed)	.090	.600	.337		.068
	Sum of Squares and Cross-products	531.601	-98.414	10.649	112.803	213.747
	Covariance	48.327	-8.947	.968	10.255	19.432
	N	12	12	12	12	12
Car value	Pearson Correlation	.913**	.052	.785**	.543	1
	Sig. (2-tailed)	<.001	.873	.002	.068	

FIGURE 2.5 Summary Statistics for Multivariate Data

From the SPSS output shown in Figure 2.5, you can interpret the different descriptive statistics, correlation and covariance values.

2.4 MULTIVARIATE PROBABILITY DISTRIBUTION

Multivariate probability distribution is used to describe the set of events (joint) of two or more random variables occurring simultaneously. The goal is to make probability statements about random vectors. The most common type of multivariate distributions is the multivariate normal distribution. Others include multinomial distribution and multivariate student's *t* distribution.

2.4.1 Multivariate Normal Distribution

The multivariate normal distribution is the most important distribution in multivariate statistics. It is the generalization of the univariate normal distribution to higher or multiple dimensions. It is described by the mean vector and the variance-covariance matrix.

Importance of multivariate normal distribution

- It is relatively easy to work with, so it is easy to obtain multivariate methods based on this particular distribution.
- Multivariate version of the central limit theorem. In the univariate course that we had a central limit theorem for the sample mean for large samples of random variables. A similar result is available in multivariate statistics that says if we have a collection of random vectors $X_1, X_2, \ldots, X_n$ that are independent and identically distributed, then the sample mean vector, $\bar{x}$, is going to be approximately multivariate normally distributed for large samples.
- Many natural phenomena may also be modeled using this distribution, just as in the univariate case.

If we have a p x 1 random vector X that is distributed according to ***a multivariate normal distribution*** with population mean vector μ and population variance-covariance matrix Σ, then this random vector, X, will have the joint density function as shown in the expression below:

$$f(x) = \left(\frac{1}{2\pi}\right)^{p/2} |\Sigma|^{-1/2} exp\left\{-\frac{1}{2}(x-\mu)' \Sigma^{-1} (x-\mu)\right\}.$$

$|\Sigma|$ denotes the determinant of the variance-covariance matrix Σ, and Σ^{-1} is just the inverse of the variance-covariance matrix Σ.

Notation

$$X \sim N(\mu, \Sigma).$$

The following term appearing inside the exponent of the multivariate normal distribution is a quadratic form:

$$(x-\mu)'\sum^{-1}(x-\mu)$$

This particular quadratic form is also called the squared ***Mahalanobis distance*** between the random vector x and the mean vector μ.

There are several techniques for assessing multivariate normal distribution. They include graphical techniques/visualizations and use of statistical tests. Both techniques can be used to gain the comprehensive understanding of your data.

One of the quickest ways to assess multivariate normality is through a probability plot: either the quantile-quantile (Q-Q) plot or the probability-probability (P-P) plot.

In SPSS, we can generate QQ plot or PP plot using the path: Analyze > Descriptive Statistics, and then selecting either the Q-Q or P-P plot > then select the multivariate variables.

Example 2.3: Using data in Figure 2.1. Assess for multivariate normality using Q-Q plot or P-P plots.

We go to: Analyze > Descriptive Statistics, and then selecting either the Q-Q plot > then select the multivariate variables (inc, age, crdbt, odbt, cvalue) (Figure 2.6).

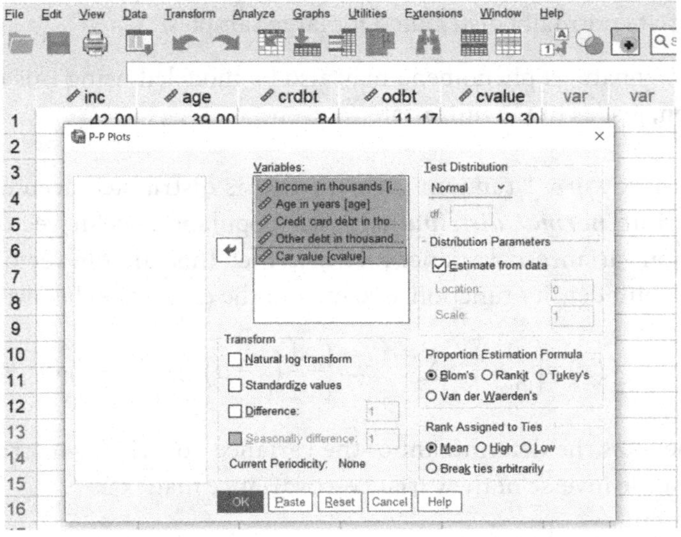

FIGURE 2.6 SPSS Procedure for Generating P-P Plot

SPSS Output will give normal P-P plots and detrended normal P-P plots for the four multivariate variables (Figure 2.7).

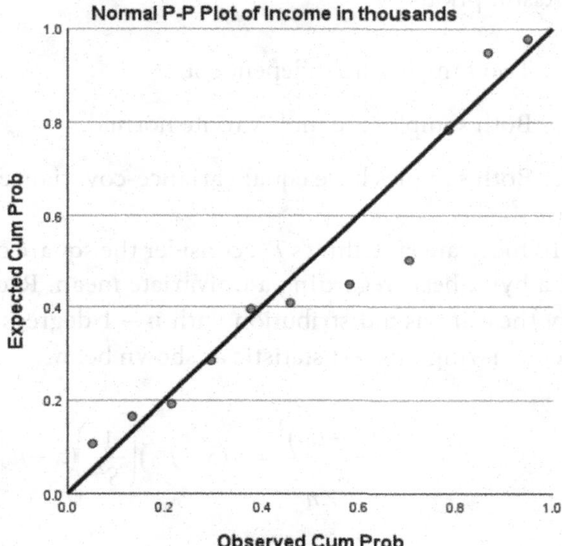

FIGURE 2.7 Normal P-P Plot

The data points are along the linear line as shown in Figure 2.7 and indicate that there is multivariate normality.

We can also use other statistical tests such as Shapiro-Wilk test, Henze-Zirkler (HZ) test, Mardia test and Doornik-Hansen test for multivariate normality or Energy test among others.

In SPSS, we can generate Shapiro-Wilk test using the path: Analyze > Descriptive Statistics > Explore > select the variable of interest and move it to Dependent List box > click Plots and select Normality plots with tests then click Continue, then OK.

2.5 INFERENCES FOR MULTIVARIATE MEANS

Inferences for multivariate means involves estimation and testing of hypothesis. Estimation is divided into two: point estimation or interval estimation. Interval estimation is also known as confidence interval estimation.

2.5.1 Multivariate One Sample Hotelling's T-Square

A more preferable test statistic is Hotelling's T^2, and we will focus on this test.

Hotelling's T-square test allows us to compare a vectors of means (for more than one variable) between one or two samples.

Assumptions:

1. Two samples are independent.
2. Both samples are multivariate normal.
3. Both samples have equal variance-covariance matrices.

To motivate Hotelling's T^2, consider the square of the t-statistic for testing a hypothesis regarding a univariate mean. Recall that under the null hypothesis t has a distribution with $n - 1$ degrees of freedom. Now consider squaring this test statistic as shown below:

$$t^2 = \frac{(\overline{x}-\mu_0)^2}{S^2/n} = n(\overline{x}-\mu_0)\left(\frac{1}{S^2}\right)(\overline{x}-\mu_0) \sim F_{1,n-1}.$$

When you square a t-distributed random variable with $n - 1$ degrees of freedom, the result is an F-distributed random variable with 1 and $n - 1$ degrees of freedom. We reject H_0 at α level if it is greater than the critical value from the F-table with 1 and $n - 1$ degrees of freedom, evaluated at α level.

$$t^2 > F_{1,n-1}.$$

Better results can be obtained from the transformation of the Hotelling T^2 statistic as below:

$$F > F_{p,n-p,\alpha}.$$

Under null hypothesis, $H_0 : \mu = \mu_0$, this will have a F distribution with p and n-p degrees of freedom. We reject the null hypothesis, H_0, at α level if the test statistic F is greater than the critical value from the F-table with p and n-p degrees of freedom, evaluated at α level.

2.5.1.1 Confidence Intervals

The *simultaneous confidence intervals* for the above are given by the expression below:

$$\bar{x}_j \pm \sqrt{\frac{p(n-1)}{(n-p)} F_{p,n-p,\alpha}} \sqrt{\frac{s_j^2}{n}}.$$

where:

$\bar{x}_j$ is the sample mean vector.
p is the number of variables (dimensions).
n is the sample size.
$F_{p,n-p,\alpha}$ is the critical value from the F distribution table with p and $n-p$ degree of freedom at α level.

2.5.1.2 Profile Plots

If the data are of a very large dimension, tables of simultaneous or Bonferroni confidence intervals are hard to grasp at a cursory glance. A better approach is to visualize the coverage of the confidence intervals through a profile plot.

2.5.2 Multivariate Paired Hotelling's T-Square

Here we are interested in testing the null hypothesis that the population mean vectors are equal against the general alternative that these mean vectors are not equal.

2.5.2.1 Confidence Intervals

Confidence intervals for the Paired Hotelling's T-square are computed in the same way as for the one-sample Hotelling's T-square. It is computed as:

$$\boldsymbol{d} \pm \sqrt{\frac{p(n-1)}{n(p-1)} F_{p,n-p,\alpha}} \sqrt{\lambda_i}.$$

where:

$\boldsymbol{d}$ is the column vector for the mean difference.
p is the number of variables (dimensions).
n is the number of pairs.
$F_{p,n-p,\alpha}$ is the critical value from the *F* distribution table with p and $n-p$ degree of freedom at α level.
λ_i is the i-th eigenvalue of the covariance matrix S_D.

2.5.3 The Two-Sample Hotelling's T-Square

It involves the computation of differences in the sample mean vectors. It also involves a calculation of the pooled variance-covariance matrix multiplied by the sum of the inverses of the sample size. The resulting matrix is then inverted.

2.5.3.1 Confidence Intervals

Calculation of simultaneous confidence intervals. Here, we calculate the Bonferroni corrected confidence intervals.

To compute Hoteling's T-square in SPSS we go to: Analyze > Scale > Reliability Analysis > select the variables of interest under Items > click Statistics and select Hoteling's T-square then select the model and the confidence level > click Continue then OK, and SPSS Output will give the Hoteling's T-square results.

Example 2.4: Consider data in Figure 2.8. Perform Hoteling's T square for two independent samples.

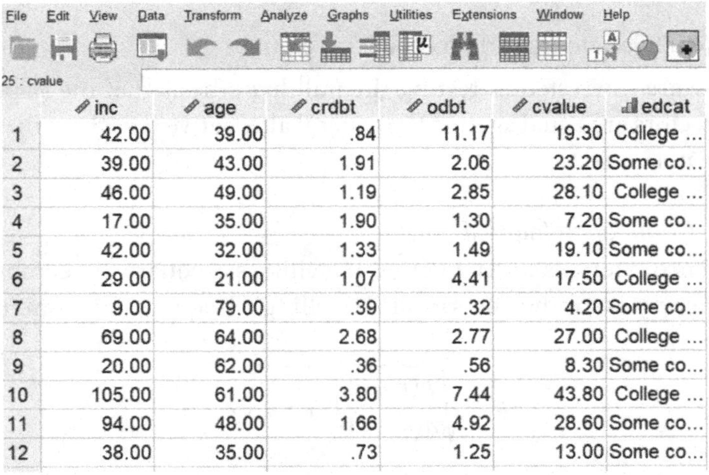

File Edit View Data Transform Analyze Graphs Utilities Extensions Window Help

25 : cvalue

	inc	age	crdbt	odbt	cvalue	edcat
1	42.00	39.00	.84	11.17	19.30	College ...
2	39.00	43.00	1.91	2.06	23.20	Some co...
3	46.00	49.00	1.19	2.85	28.10	College ...
4	17.00	35.00	1.90	1.30	7.20	Some co...
5	42.00	32.00	1.33	1.49	19.10	Some co...
6	29.00	21.00	1.07	4.41	17.50	College ...
7	9.00	79.00	.39	.32	4.20	Some co...
8	69.00	64.00	2.68	2.77	27.00	College ...
9	20.00	62.00	.36	.56	8.30	Some co...
10	105.00	61.00	3.80	7.44	43.80	College ...
11	94.00	48.00	1.66	4.92	28.60	Some co...
12	38.00	35.00	.73	1.25	13.00	Some co...

FIGURE 2.8 Sample Multivariate Data

We go to: Analyze > General Linear Model > Multivariate > select the multivariate numeric variables as Dependent variables and the categorical variable (level of education) as the factor variable > click Options and select descriptives and homogeneity tests (Figure 2.9).

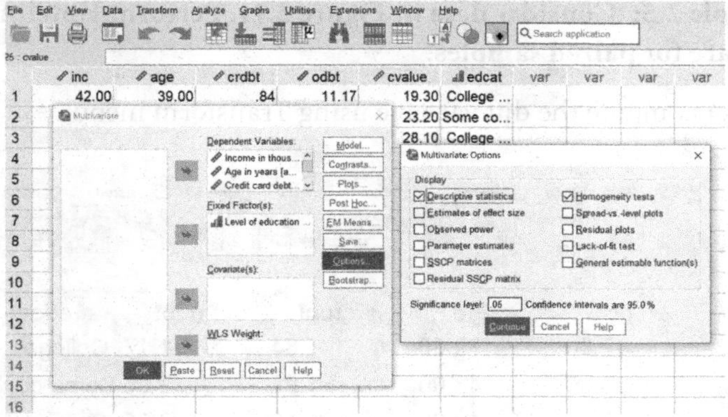

FIGURE 2.9 SPSS Procedure for Hoteling's T Square for Independent Samples

SPSS Output will give the Hoteling's T Square test for two independent samples as shown in Figure 2.10.

Descriptive Statistics

	Level of education	Mean	Std. Deviation	N
Income in thousands	Some college	37.0000	28.14249	7
	College degree	58.2000	29.87809	5
	Total	45.8333	29.59371	12
Age in years	Some college	47.7143	17.20188	7
	College degree	46.8000	17.52712	5
	Total	47.3333	16.53280	12
Credit card debt in thousands	Some college	1.1817	.68431	7
	College degree	1.9166	1.27683	5
	Total	1.4879	.99572	12
Other debt in thousands	Some college	1.6991	1.53536	7
	College degree	5.7288	3.58112	5
	Total	3.3781	3.20232	12
Car value	Some college	14.8000	9.08387	7
	College degree	27.1400	10.40303	5
	Total	19.9417	11.16866	12

Multivariate Tests[a]

Effect		Value	F	Hypothesis df	Error df	Sig.
Intercept	Pillai's Trace	.956	25.851[b]	5.000	6.000	<.001
	Wilks' Lambda	.044	25.851[b]	5.000	6.000	<.001
	Hotelling's Trace	21.542	25.851[b]	5.000	6.000	<.001
	Roy's Largest Root	21.542	25.851[b]	5.000	6.000	<.001
edcat	Pillai's Trace	.653	2.262[b]	5.000	6.000	.175
	Wilks' Lambda	.347	2.262[b]	5.000	6.000	.175
	Hotelling's Trace	1.885	2.262[b]	5.000	6.000	.175
	Roy's Largest Root	1.885	2.262[b]	5.000	6.000	.175

a. Design: Intercept + edcat

b. Exact statistic

FIGURE 2.10 SPSS Output for Hoteling's T Square Test for Two Independent Samples

Example 2.5: Consider data in Figure 2.11. Perform Hoteling's T square for paired samples.

We first compute the diff variable using Transform menu.

File Edit View Data Transform Analyze Graphs Utilities Extensions Window

	inc	age	crdbt	odbt	edcat
1	42.00	39.00	.84	11.17	College ...
2	39.00	43.00	1.91	2.06	Some co...
3	46.00	49.00	1.19	2.85	College ...
4	17.00	35.00	1.90	1.30	Some co...
5	42.00	32.00	1.33	1.49	Some co...
6	29.00	21.00	1.07	4.41	College ...
7	9.00	79.00	.39	.32	Some co...
8	69.00	64.00	2.68	2.77	College ...
9	20.00	62.00	.36	.56	Some co...
10	105.00	61.00	3.80	7.44	College ...
11	94.00	48.00	1.66	4.92	Some co...
12	38.00	35.00	.73	1.25	Some co...

FIGURE 2.11 Multivariate Data for Paired Hoteling's T Square

We go to: Analyze > General Linear Model > Multivariate > select the difference in multivariate numeric variables as dependent variables and the categorical variable (level of education) as the factor variable > click Options and select descriptives and homogeneity tests. SPSS Output will give the Hoteling's T Square test for paired samples.

2.5.4 Multivariate Analysis of Variance (MANOVA)

The multivariate analysis of variance (MANOVA) is the multivariate analog of the analysis of variance (ANOVA) procedure used for univariate data. MANOVA also investigates the effects of a categorical variable (the groups, i.e., independent variables) on a continuous outcome, but in this case, the outcome is represented by a vector of dependent variables.

2.5.4.1 Assumptions

The assumptions here are essentially the same as the assumptions in a Hotelling's test, only here they apply to groups:

- The data from group i has common mean vector.
- The data from all groups have common variance-covariance matrix.
- Independence: The subjects are independently sampled.
- Normality: The data are multivariate normally distributed.

2.5.4.2 Test Statistics for MANOVA

Let **H** be the hypothesis sums of squares and cross products matrix, and let the error sums of squares and cross products matrix be represented be **E**. The multivariate equivalent of the **A** statistic is the matrix **A** given by:

$$\boldsymbol{A} = \boldsymbol{H}\boldsymbol{E}^{-1}.$$

Let there be q dependent variables in the MANOVA, and let λ_i denote the ith eigenvalue of matrix A.

There are four different test statistics based on the MANOVA table:

1. **Wilks Lambda**

   Here, the determinant of the error sums of squares and cross products matrix **E** is divided by the determinant of the total sum of squares and cross products matrix **T** = **H** + **E**. If **H** is large relative to **E**, then |**H** + **E**| will be large relative to |**E**|. Thus, we will reject the null hypothesis if Wilks lambda is small (close to zero).

$$\text{Wilk's lambda} = \Lambda = \frac{|E|}{|H+E|} = \prod_{i=1}^{q} \frac{1}{1+\lambda_i}.$$

2. **Hotelling-Lawley trace**

   Here, we are multiplying **H** by the inverse of **E**; then we take the trace of the resulting matrix. If **H** is large relative to **E**, then the Hotelling-Lawley trace will take a large value. Thus, we will reject the null hypothesis if this test statistic is large.

$$\text{Hotelling–Lawley's trace} = \text{trace}(\text{A}) = \text{trace}\left(\boldsymbol{H}\boldsymbol{E}^{-1}\right) = \sum_{i=1}^{q} \lambda_i .$$

3. **Pillai trace**

   Here, we are multiplying **H** by the inverse of the total sum of squares and cross products matrix **T** = **H** + **E**. If **H** is large relative to **E**, then the Pillai trace will take a large value. Thus, we will reject the null hypothesis if this test statistic is large.

$$\text{Pillai's trace} = \text{trace}\left[H(H+E)^{-1}\right] = \sum_{i=1}^{q} \frac{\lambda_i}{1+\lambda_i}.$$

4. **Roy's maximum root: Largest eigenvalue of HE–[1]**

   Here, we multiply **H** by the inverse of **E** and then compute the largest eigenvalue of the resulting matrix. If **H** is large relative to **E**, then the Roy's root will take a large value. Thus, we will reject the null hypothesis if this test statistic is large.

$$\text{Roy's largest root} = \max(\lambda_i)$$

To compute MANOVA in SPSS, we go to: Analyze > General Linear Model > Multivariate > select the several continuous variables under Dependent Variables and the categorical variable of interest under Fixed Factor(s) > click Post Hoc test and select Bonferroni then click Continue then OK. SPSS Output will give the MANOVA results.

Example 2.6: Consider data in Figure 2.12. Perform MANOVA.

File Edit View Data Transform Analyze Graphs Utilities Extensions Window Help

	inc	age	crdbt	odbt	cvalue	edcat
1	42.00	39.00	.84	11.17	19.30	College ...
2	39.00	43.00	1.91	2.06	23.20	Some co...
3	46.00	49.00	1.19	2.85	28.10	High sch...
4	17.00	35.00	1.90	1.30	7.20	Did not c...
5	42.00	32.00	1.33	1.49	19.10	Some co...
6	29.00	21.00	1.07	4.41	17.50	High sch...
7	9.00	79.00	.39	.32	4.20	High sch...
8	69.00	64.00	2.68	2.77	27.00	Did not c...
9	20.00	62.00	.36	.56	8.30	Some co...
10	105.00	61.00	3.80	7.44	43.80	Did not c...
11	94.00	48.00	1.66	4.92	28.60	College ...
12	38.00	35.00	.73	1.25	13.00	Some co...

FIGURE 2.12 MANOVA Data

We go to: Analyze > General Linear Model > Multivariate > select the five numeric variables (inc, age, crdbt, odbt, cvalue) under Dependent Variables and the categorical variable (level of education) under Fixed Factor(s) > click Post Hoc then drag the Factor variable (edcat) to Post Hoc Tests for: and select Bonferroni then click Continue then OK (Figure 2.13).

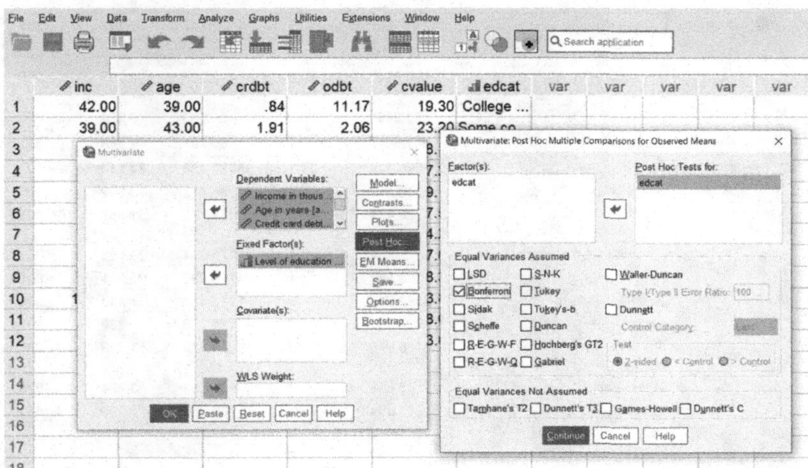

FIGURE 2.13 SPSS procedure for MANOVA

SPSS Output will give the MANOVA results displayed in Figure 2.14.

Multivariate Tests[a]

Effect		Value	F	Hypothesis df	Error df	Sig.
Intercept	Pillai's Trace	.962	20.025[b]	5.000	4.000	.006
	Wilks' Lambda	.038	20.025[b]	5.000	4.000	.006
	Hotelling's Trace	25.031	20.025[b]	5.000	4.000	.006
	Roy's Largest Root	25.031	20.025[b]	5.000	4.000	.006
edcat	Pillai's Trace	1.889	2.041	15.000	18.000	.075
	Wilks' Lambda	.016	2.668	15.000	11.444	.050
	Hotelling's Trace	13.140	2.336	15.000	8.000	.114
	Roy's Largest Root	8.278	9.934[c]	5.000	6.000	.007

a. Design: Intercept + edcat

b. Exact statistic

c. The statistic is an upper bound on F that yields a lower bound on the significance level.

FIGURE 2.14 Additional MANOVA Results

There will be additional outputs for the MANOVA shown in Figure 2.15 (test of between subjects effects) and Figure 2.16 (multiple comparisons), respectively.

Tests of Between-Subjects Effects

Source	Dependent Variable	Type III Sum of Squares	df	Mean Square	F	Sig.
Corrected Model	Income in thousands	3382.250[a]	3	1127.417	1.443	.301
	Age in years	228.833[b]	3	76.278	.220	.880
	Credit card debt in thousands	6.976[c]	3	2.325	4.733	.035
	Other debt in thousands	63.008[d]	3	21.003	3.374	.075
	Car value	241.084[e]	3	80.361	.568	.651
Intercept	Income in thousands	26680.828	1	26680.828	34.144	<.001
	Age in years	25348.412	1	25348.412	73.002	<.001
	Credit card debt in thousands	25.475	1	25.475	51.857	<.001
	Other debt in thousands	175.064	1	175.064	28.126	<.001
	Car value	4798.590	1	4798.590	33.941	<.001
edcat	Income in thousands	3382.250	3	1127.417	1.443	.301
	Age in years	228.833	3	76.278	.220	.880
	Credit card debt in thousands	6.976	3	2.325	4.733	.035
	Other debt in thousands	63.008	3	21.003	3.374	.075
	Car value	241.084	3	80.361	.568	.651
Error	Income in thousands	6251.417	8	781.427		
	Age in years	2777.833	8	347.229		
	Credit card debt in thousands	3.930	8	.491		
	Other debt in thousands	49.795	8	6.224		
	Car value	1131.045	8	141.381		
Total	Income in thousands	34842.000	12			
	Age in years	29892.000	12			
	Credit card debt in thousands	37.474	12			
	Other debt in thousands	249.745	12			
	Car value	6144.170	12			
Corrected Total	Income in thousands	9633.667	11			
	Age in years	3006.667	11			
	Credit card debt in thousands	10.906	11			
	Other debt in thousands	112.803	11			
	Car value	1372.129	11			

a. R Squared = .351 (Adjusted R Squared = .108)
b. R Squared = .076 (Adjusted R Squared = -.270)
c. R Squared = .640 (Adjusted R Squared = .505)
d. R Squared = .559 (Adjusted R Squared = .393)
e. R Squared = .176 (Adjusted R Squared = -.133)

FIGURE 2.15

Multiple Comparisons

Bonferroni

Dependent Variable	(I) Level of education	(J) Level of education	Mean Difference (I-J)	Std. Error	Sig.	95% Confidence Interval Lower Bound	Upper Bound
Income in thousands	Did not complete high school	High school degree	35.6667	22.82436	.941	-43.7365	115.0699
		Some college	28.9167	21.35023	1.000	-45.3582	103.1915
		College degree	-4.3333	25.51841	1.000	-93.1088	84.4421
	High school degree	Did not complete high school	-35.6667	22.82436	.941	-115.0699	43.7365
		Some college	-6.7500	21.35023	1.000	-81.0249	67.5249
		College degree	-40.0000	25.51841	.934	-128.7755	48.7755
	Some college	Did not complete high school	-28.9167	21.35023	1.000	-103.1915	45.3582
		High school degree	6.7500	21.35023	1.000	-67.5249	81.0249
		College degree	-33.2500	24.20889	1.000	-117.4698	50.9698
	College degree	Did not complete high school	4.3333	25.51841	1.000	-84.4421	93.1088
		High school degree	40.0000	25.51841	.934	-48.7755	128.7755
		Some college	33.2500	24.20889	1.000	-50.9698	117.4698
Age in years	Did not complete high school	High school degree	3.6667	15.21467	1.000	-49.2633	56.5967
		Some college	10.3333	14.23202	1.000	-39.1781	59.8448
		College degree	9.8333	17.01052	1.000	-49.3442	69.0109
	High school degree	Did not complete high school	-3.6667	15.21467	1.000	-56.5967	49.2633
		Some college	6.6667	14.23202	1.000	-42.8448	56.1781
		College degree	6.1667	17.01052	1.000	-53.0109	65.3442
	Some college	Did not complete high school	-10.3333	14.23202	1.000	-59.8448	39.1781
		High school degree	-6.6667	14.23202	1.000	-56.1781	42.8448
		College degree	-.5000	16.13759	1.000	-56.6407	55.6407
	College degree	Did not complete high school	-9.8333	17.01052	1.000	-69.0109	49.3442
		High school degree	-6.1667	17.01052	1.000	-65.3442	53.0109
		Some college	.5000	16.13759	1.000	-55.6407	56.6407
Credit card debt in thousands	Did not complete high school	High school degree	1.9098	.57228	.062	-.0811	3.9007
		Some college	1.7100	.53532	.076	-.1523	3.5723
		College degree	1.5431	.63983	.254	-.6828	3.7690
	High school degree	Did not complete high school	-1.9098	.57228	.062	-3.9007	.0811
		Some college	-.1998	.53532	1.000	-2.0621	1.6625
		College degree	-.3667	.63983	1.000	-2.5926	1.8592
	Some college	Did not complete high school	-1.7100	.53532	.076	-3.5723	.1523
		High school degree	.1998	.53532	1.000	-1.6625	2.0621
		College degree	-.1669	.60699	1.000	-2.2786	1.9447
	College degree	Did not complete high school	-1.5431	.63983	.254	-3.7690	.6828
		High school degree	.3667	.63983	1.000	-1.8592	2.5926
		Some college	.1669	.60699	1.000	-1.9447	2.2786
Other debt in thousands	Did not complete high school	High school degree	1.3072	2.03705	1.000	-5.7794	8.3938
		Some college	2.4953	1.90548	1.000	-4.1336	9.1243
		College degree	-4.2117	2.27749	.610	-12.1349	3.7114
	High school degree	Did not complete high school	-1.3072	2.03705	1.000	-8.3938	5.7794
		Some college	1.1881	1.90548	1.000	-5.4408	7.8171
		College degree	-5.5190	2.27749	.250	-13.4421	2.4042
	Some college	Did not complete high school	-2.4953	1.90548	1.000	-9.1243	4.1336
		High school degree	-1.1881	1.90548	1.000	-7.8171	5.4408
		College degree	-6.7071	2.16062	.087	-14.2236	.8095
	College degree	Did not complete high school	4.2117	2.27749	.610	-3.7114	12.1349
		High school degree	5.5190	2.27749	.250	-2.4042	13.4421
		Some college	6.7071	2.16062	.087	-.8095	14.2236
Car value	Did not complete high school	High school degree	9.4000	9.70844	1.000	-24.3745	43.1745
		Some college	10.1000	9.08141	1.000	-21.4931	41.6931
		College degree	2.0500	10.85436	1.000	-35.7110	39.8110
	High school degree	Did not complete high school	-9.4000	9.70844	1.000	-43.1745	24.3745
		Some college	.7000	9.08141	1.000	-30.8931	32.2931
		College degree	-7.3500	10.85436	1.000	-45.1110	30.4110
	Some college	Did not complete high school	-10.1000	9.08141	1.000	-41.6931	21.4931
		High school degree	-.7000	9.08141	1.000	-32.2931	30.8931
		College degree	-8.0500	10.29735	1.000	-43.8732	27.7732
	College degree	Did not complete high school	-2.0500	10.85436	1.000	-39.8110	35.7110
		High school degree	7.3500	10.85436	1.000	-30.4110	45.1110
		Some college	8.0500	10.29735	1.000	-27.7732	43.8732

Based on observed means.
The error term is Mean Square(Error) = 141.381.

FIGURE 2.16

Wilks' Lambda is the statistic of choice for many, and it's the one that is usually reported.

2.6 PRINCIPAL COMPONENT ANALYSIS

Principal components analysis is a way of re-describing the variation observed in your data. It serves as a means of reducing the dimensionality of data (i.e. reducing the number of predictor variables) and is often used for exploratory analyses.

The purpose of principal component analysis is to find the best low-dimensional representation of the variation in a multivariate data set. For example, in the case of the data set with 13 variables describing the data. We can conduct principal component analysis to investigate whether we can capture most of the variation between samples using a smaller number of new variables (principal components), where each of these new variables is a linear combination of all or some of the 15 variables.

Let $\vec{X} = \begin{bmatrix} X_1 \\ \vdots \\ X_p \end{bmatrix}$ be a random vector with population mean $\vec{\mu}$ and population covariance matrix Σ.

The spectral decomposition of Σ is given by

$$\sum = \lambda_1 \vec{\upsilon}_1 \vec{\upsilon}_1^T + \ldots + \lambda_p \vec{\upsilon}_p \vec{\upsilon}_p^T.$$

where $\lambda_1 > \lambda_2 > \ldots > \lambda_p > 0$ and each eigen vector is represented by

$$\vec{\upsilon}_k = \left[\upsilon_{k1}, \ldots \upsilon_{kp}\right]^T.$$

The first principal component

Consider a linear combination of the variates by $\vec{a} = [a_1, \ldots a_p]^T$:

$$Y_1 = \vec{a}^T \vec{X} = a_1 X_1 + a_2 X_2 + \ldots + a_p X_p.$$

In order to explain the variance-covariance of $\vec{X}$ as much as possible, we want to maximize the variance of Y_1. At the same time, in order to fix the scale, we impose the constraint $\| a \| = 1$. Then, the first principal component for $\vec{X}$ is defined by the following optimization:

$$\max Var(Y_1) = Var\left(\vec{a}^T \vec{X}\right) = \vec{a}^T \sum \vec{a}.$$

$$s.t \ \|a\| = 1.$$

The Lagrangian for the above optimization is

$$f(\vec{a};\lambda)=\vec{a}^T\sum\vec{a}-\lambda\left(\vec{a}^T\vec{a}-1\right).$$

and by setting the gradient to zero

$$\sum\vec{a}=\lambda\vec{a}.$$

This implies that $\vec{a}$ must be a unit eigenvector, and λ is the corresponding eigenvalue. Notice what we aim to maximize is

$$Var(Y_1)=Var\left(\vec{a}^T\vec{X}\right)=\vec{a}^T\sum\vec{a}=Var\left(\vec{a}^T\vec{X}\right)=\lambda\vec{a}^T\vec{a}=\lambda.$$

since $\lambda_1>\lambda_2>\ldots>\lambda_p>0$.

Therefore, the first principal component is

$$Y_1=\vec{\upsilon}_1^T\vec{X}=\upsilon_{11}X_1+\upsilon_{12}X_2+\ldots+\upsilon_{1p}X_p.$$

where $\vec{\upsilon}_1$ is the eigenvector corresponding to the leading eigenvalue λ_1. Moreover, $Var(Y_1)=\lambda_1$.

The second principal component is

$$Y_2=\vec{\upsilon}_2^T\vec{X}=\upsilon_{21}X_1+\upsilon_{22}X_2+\ldots+\upsilon_{2p}X_p.$$

where $\vec{\upsilon}_2$ is the eigenvector corresponding to the leading eigenvalue λ_2. Moreover, $Var(Y_2)=\lambda_2$.

The kth principal component is

$$Y_k=\vec{\upsilon}_k^T\vec{X}=\upsilon_{k1}X_1+\upsilon_{k2}X_2+\ldots+\upsilon_{kp}X_p.$$

where $\vec{\upsilon}_k$ is the eigenvector corresponding to the leading eigenvalue λ_k. Moreover, $Var(Y_k)=\lambda_k$.

The coefficients υ_{k1}, υ_{k2}, ..., υ_{kp} are referred to as loadings on the random variables X_1, ..., X_p for the kth principal component Y_k.

2.6.1 Standardization

In certain applications, it is common to standardize the original variates $X_1, \ldots, X_p$ into

$$Z_1=\frac{X_1-\mu_1}{\sqrt{\sigma_{11}}},\ Z_2=\frac{X_2-\mu_2}{\sqrt{\sigma_{22}}},\ldots,\ Z_p=\frac{X_p-\mu_p}{\sqrt{\sigma_{pp}}}.$$

Selecting the number of PCs.

The proportion of the total variance due to the first k principal components is defined as

$$\frac{Var(Y_1)+\ldots+Var(Y_k)}{Var(Y_1)+\ldots+Var(Y_p)}=\frac{\lambda_1+\ldots+\lambda_k}{\lambda_1+\ldots+\lambda_p}=\frac{\lambda_1+\ldots+\lambda_k}{\sigma_{11}+\ldots+\sigma_{pp}}.$$

2.6.2 Standardizing Variables

If you want to compare different variables that have different units and are very different variances, it is a good idea to first standardize the variables.

To perform principal component analysis (PCA) on a multivariate data set, the first step is often to standardize the variables under consideration.

Once you have standardized your variables, you can perform principal component analysis.

2.6.3 Deciding How Many Principal Components to Retain

In order to decide how many principal components should be retained, it is common to summarize the results of a principal components analysis by making a scree plot.

Another way of deciding how many components to retain is to use Kaiser's criterion: that we should only retain principal components for which the variance is above 1 (when principal component analysis was applied to standardized data). We can check this by finding the variance of each of the principal components.

A third way to decide how many principal components to retain is to decide to keep the number of components required to explain at least some minimum amount of the total variance.

2.6.4 Loadings for the Principal Components

The loadings for the principal components are stored in a named element "rotation" of the variable. This contains a matrix with the loadings of each principal component, where the first column in the matrix contains the loadings for the first principal component, the second column contains the loadings for the second principal component and so on.

2.6.5 Scatterplots of the Principal Components

The values of the principal components are stored in a named element "x" of the variable. This contains a matrix with the principal components, where the first column in the matrix contains the first principal component, the second column the second component and so on.

To perform principal component analysis in SPSS, we go to: Analyze > Dimension Reduction > Factor > Transfer all the variables you want include in the analysis into the Variables: box > Click on the Descriptives and under Statistics area, select Initial solution and under Correlation Matrix area, select Coefficients, KMO and Bartlett's test of sphericity, Reproduced and Anti-image and click Continue > click Extraction and keep all the defaults and also select Scree plot under Display then click Continue > click Rotation and Select the Varimax option under Method. This will activate the Rotated solution option in the Display area and will be checked by default and also select Loading plot(s) in the Display area then click Continue > click Scores and check the Save as variables option and then keep the Regression option selected then click Continue > click Options then check the Sorted by size and Suppress small coefficients option. Change the Absolute value below: from ".10" to ".3" then click Continue then OK. SPSS Output will generate the results for Principals Component Analysis.

Example 2.7: Consider the following data on a five-point Likert scale (Figure 2.17) with four variables for job security (q6), career development (q7), job flexibility (q8) and employee retention (q9). Perform PCA on job security.

PCA data.sav [DataSet6] - IBM SPSS Statistics Data Editor

File Edit View Data Transform Analyze Graphs Utilities Extensions Window Help

Search application

	q6a	q6b	q6c	q6d	q6e	q6f	q7a	q7b	q7c	q7d
1	Agree	Strongly ...	Strongly ...	Strongly ...	Neutral	Agree	Neutral	Neutral	Agree	Neutral
2	Neutral	Strongly A...	Agree	Agree	Disagree	Agree	Agree	Agree	Agree	Agree
3	Strongly A...	Disagree	Disagree	Neutral	Agree	Agree	Agree	Agree	Agree	Agree
4	Strongly A...	Agree	Disagree	Disagree	Neutral	Strongly A...	Neutral	Strongly A...	Strongly A...	Neutral
5	Agree	Agree	Neutral	Agree	Strongly ...	Strongly A...	Agree	Agree	Agree	Agree
6	Strongly A...	Strongly A...	Strongly A...	Disagree	Neutral	Strongly A...	Neutral	Strongly A...	Strongly A...	Neutral
7	Agree	Strongly A...	Disagree	Disagree	Disagree	Strongly A...	Agree	Agree	Agree	Disagree
8	Neutral	Disagree	Neutral	.	Neutral	Disagree	Neutral	Neutral	Disagree	Neutral
9	Neutral	Agree	Neutral	Disagree	Neutral	Strongly A...	Agree	Neutral	Agree	Disagree
10										

FIGURE 2.17 A Five-Point Likert Scale Data

We go to: Analyze > Dimension Reduction > Factor > Transfer all the variables describing job security (q6a to q6e) into the Variables: box > Click on the Descriptives and under Statistics area, select Initial solution and under Correlation Matrix area select Coefficients, KMO and Bartlett's test of sphericity, Reproduced and Anti-image and click Continue > click Extraction and keep all the defaults and also select Scree plot under Display then click Continue > click Rotation and Select the Varimax option under Method. This will activate the Rotated solution option in the Display area and will be checked by default and also select Loading plot(s) in the Display area then click Continue > click Scores and check the Save as variables option and then keep the Regression option selected then click Continue > click Options then check the Sorted by size and Suppress small coefficients option. Change the Absolute value below: from ".10" to ".3" then click Continue then OK (Figure 2.18).

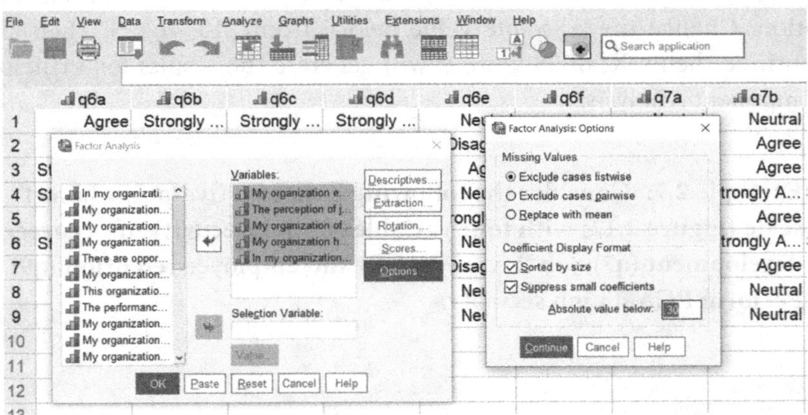

FIGURE 2.18 SPSS Procedure for PCA

SPSS Output will generate the results for Principal Component Analysis, which includes Correlation Matrix and KMO (Figure 2.19), Anti-Image Matrices (Figure 2.20), communalities and Total variance Explained (Figure 2.21), Scree plot (Figure 2.22), Component Matrix (Figure 2.23) and Rotated Component matrix.

Correlation Matrix

		My organization ensures that employees earn fair compensation	The perception of job security positively affects retention	My organization offers long term contracts	My organization has long term funding	In my organization, your employment can be terminated at any time
Correlation	My organization ensures that employees earn fair compensation	1.000	-.201	-.100	-.240	.444
	The perception of job security positively affects retention	-.201	1.000	.712	.359	-.498
	My organization offers long term contracts	-.100	.712	1.000	.417	-.213
	My organization has long term funding	-.240	.359	.417	1.000	-.511
	In my organization, your employment can be terminated at any time	.444	-.498	-.213	-.511	1.000

KMO and Bartlett's Test

Kaiser-Meyer-Olkin Measure of Sampling Adequacy.		.540
Bartlett's Test of Sphericity	Approx. Chi-Square	7.871
	df	10
	Sig.	.641

FIGURE 2.19 SPSS Output for the Correlation Matrix and KMO

Anti-image Matrices

		My organization ensures that employees earn fair compensation	The perception of job security positively affects retention	My organization offers long term contracts	My organization has long term funding	In my organization, your employment can be terminated at any time
Anti-image Covariance	My organization ensures that employees earn fair compensation	.802	-.021	.016	.006	-.216
	The perception of job security positively affects retention	-.021	.355	-.270	.087	.206
	My organization offers long term contracts	.016	-.270	.398	-.190	-.157
	My organization has long term funding	.006	.087	-.190	.617	.248
	In my organization, your employment can be terminated at any time	-.216	.206	-.157	.248	.470
Anti-image Correlation	My organization ensures that employees earn fair compensation	.706[a]	-.040	.029	.009	-.353
	The perception of job security positively affects retention	-.040	.533[a]	-.720	.186	.503
	My organization offers long term contracts	.029	-.720	.480[a]	-.384	-.362
	My organization has long term funding	.009	.186	-.384	.612[a]	.460
	In my organization, your employment can be terminated at any time	-.353	.503	-.362	.460	.510[a]

a. Measures of Sampling Adequacy(MSA)

FIGURE 2.20 SPSS Output for Anti-Image Matrices

Communalities

	Initial	Extraction
My organization ensures that employees earn fair compensation	1.000	.710
The perception of job security positively affects retention	1.000	.797
My organization offers long term contracts	1.000	.852
My organization has long term funding	1.000	.527
In my organization, your employment can be terminated at any time	1.000	.742

Extraction Method: Principal Component Analysis.

Total Variance Explained

Component	Initial Eigenvalues			Extraction Sums of Squared Loadings			Rotation Sums of Squared Loadings		
	Total	% of Variance	Cumulative %	Total	% of Variance	Cumulative %	Total	% of Variance	Cumulative %
1	2.520	50.404	50.404	2.520	50.404	50.404	1.962	39.242	39.242
2	1.108	22.158	72.562	1.108	22.158	72.562	1.666	33.320	72.562
3	.678	13.568	86.130						
4	.515	10.306	96.436						
5	.178	3.564	100.000						

Extraction Method: Principal Component Analysis.

FIGURE 2.21 SPSS Output for the Communalities and Total Variance Explained

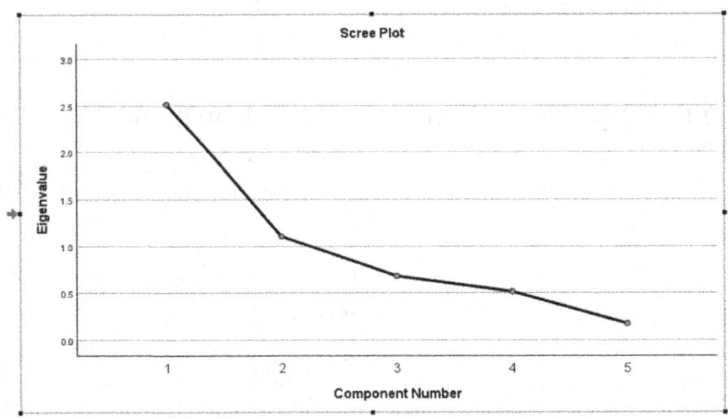

FIGURE 2.22 SPSS Output for the Scree Plot

Component Matrix[a]

	Component	
	1	2
The perception of job security positively affects retention	.819	.354
In my organization, your employment can be terminated at any time	-.755	.415
My organization has long term funding	.722	
My organization offers long term contracts	.720	.578
My organization ensures that employees earn fair compensation	-.490	.685

Extraction Method: Principal Component Analysis.

a. 2 components extracted.

FIGURE 2.23 SPSS Output for the Component Matrix

2.7 FACTOR ANALYSIS

Factor analysis is a way to condense the data in many variables into just a few variables. For this reason, it is also known as "dimension reduction". It makes the grouping of variables with high correlation. Factor analysis includes techniques such as principal component analysis and common factor analysis.

Factor analysis is totally dependent on correlations between variables where we summarize the correlation structure.

This type of technique is used as a preprocessing step to transform the data before using other models. When the data have too many variables, the performance of multivariate techniques is not at the optimum level, as patterns are more difficult to find. By using factor analysis, the patterns become less diluted and easier to analyze.

Exploratory factor analysis (EFA) should be used when you need to develop a hypothesis about relationship between variables. Confirmatory factor analysis (CFA) should be used when you need to test a hypothesis about relationship between variables.

Example of Exploratory Factor Analysis considering one common factor model

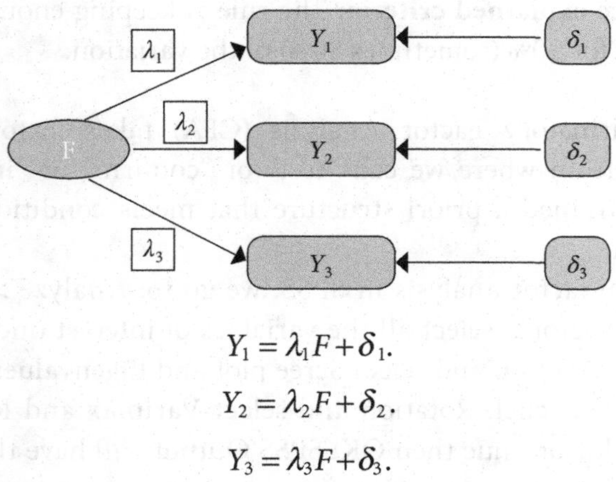

$$Y_1 = \lambda_1 F + \delta_1.$$

$$Y_2 = \lambda_2 F + \delta_2.$$

$$Y_3 = \lambda_3 F + \delta_3.$$

The factor F is not observed; only Y_1, Y_2, Y_3 are observed:

δ_i represent variability in the Y_i NOT explained by F.

Y_i is a linear function of F and δ_i.

Given all variables in standardized from, i.e., $Var(Y_i) = Var(F) = 1$.

The factor loadings is given by $\lambda_i = Corr(Y_i, F)$.

The communality of Y_i is given by $h_i^2 = [Corr(Y_i, F)]^2$ which is the % variance of Y_i explained by F.

The uniqueness of Y_i is given by $1 - h_i^2$ which is the residual variance of Y_i.

The degree of factorial determination is given by $\Sigma\lambda_1^2/n$ where n is the number of observed variable Y.

To extract the initial factors, we can use Least-squares method (e.g., principal axis factoring with iterated communalities) or the Maximum likelihood method.

There are several criteria for determining the number of factors. They include the following:

1. **Comprehensibility:** This is where one limits the number of factors this is the first two or three.
2. **Kaiser criterion:** The Kaiser rule is to drop all components with eigenvalues under 1.0.
3. **Scree plot:** Drop all further components after the one starting the elbow.
4. **Variance explained criteria:** The rule is keeping enough factors to account for 90% (sometimes 80%) of the variation.

The Confirmatory Factor Analysis (CFA) takes factor analysis a step further from where we can "test" or "confirm" or "implement" a "highly constrained a priori structure that meets conditions of model identification."

To perform factor analysis in SPSS, we go to: Analyze > Dimension Reduction > Factor > select all the variables of interest under variables: box > click Extraction and select Scree plot and Eigenvalues over 1 then click Continue > click Rotation and select Varimax and Rotated solution then click Continue then OK. SPSS Output will have the results for Factor Analysis.

Example 2.8: Consider the following data on a five-point Likert scale (Figure 2.24) with four variables for job security (q6), career development (q7), job flexibility (q8) and employee retention (q9). Perform factor analysis on job security.

Rotated Component Matrix[a]

	Component 1	Component 2
My organization offers long term contracts	.923	
The perception of job security positively affects retention	.860	
My organization ensures that employees earn fair compensation		.841
In my organization, your employment can be terminated at any time	-.326	.797
My organization has long term funding	.512	-.515

Extraction Method: Principal Component Analysis.
Rotation Method: Varimax with Kaiser Normalization.

a. Rotation converged in 3 iterations.

FIGURE 2.24 SPSS Output for Rotated Component Matrix

We go to: Analyze > Dimension Reduction > Factor > select all the variables of interest (q6) under variables: box > click Extraction and select Scree plot and Eigenvalues over 1 then click Continue > click Rotation and select Varimax and Rotated solution then click Continue then OK (Figure 2.25).

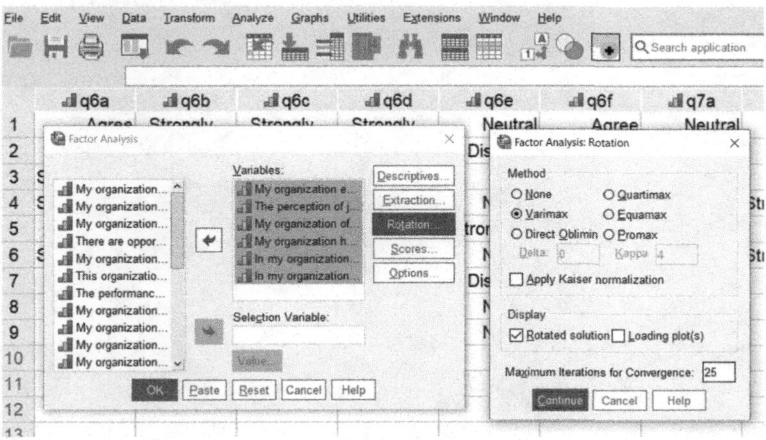

FIGURE 2.25 SPSS Procedure for Factor Analysis

SPSS Output will have the results for factor analysis – total variance explained and scree plot (Figure 2.26) plus Rotated component Matrix (Figure 2.27).

Total Variance Explained

Component	Initial Eigenvalues			Rotation Sums of Squared Loadings		
	Total	% of Variance	Cumulative %	Total	% of Variance	Cumulative %
1	2.684	44.731	44.731	2.162	36.029	36.029
2	1.395	23.257	67.988	1.918	31.959	67.988
3	.937	15.618	83.607			
4	.644	10.739	94.345			
5	.190	3.167	97.512			
6	.149	2.488	100.000			

Extraction Method: Principal Component Analysis.

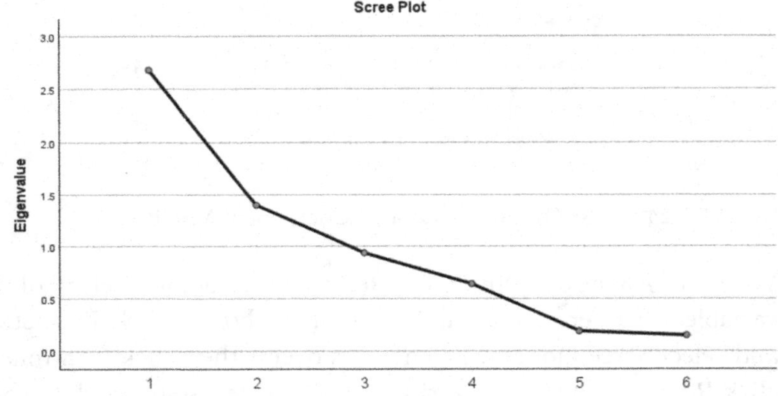

FIGURE 2.26 SPSS Output for Total Variance Explained and Scree Plot

Rotated Component Matrix[a]

	Component 1	Component 2
My organization ensures that employees earn fair compensation	.103	.749
The perception of job security positively affects retention	.880	-.334
My organization offers long term contracts	.682	-.318
My organization has long term funding	.151	-.784
In my organization, your employment can be terminated at any time	-.402	.690
In my organization, there is a clear process for termination	.852	.232

Extraction Method: Principal Component Analysis.
Rotation Method: Varimax without Kaiser Normalization.

a. Rotation converged in 3 iterations.

Component Transformation Matrix

Component	1	2
1	.771	-.637
2	.637	.771

Extraction Method: Principal Component Analysis.
Rotation Method: Varimax without Kaiser Normalization.

FIGURE 2.27 SPSS Output for Rotated Component Matrix

2.8 STATISTICAL LEARNING

Statistical learning can be categorized into supervised and unsupervised learning. In unsupervised learning, we have clustering techniques where learning is without labels i.e. finding subtypes or groups that are not defined a priori based on measurements, while in the supervised learning we have classification techniques where we use a priori group labels in analysis to assign new observations to a particular group or class.

2.8.1 Cluster Analysis

Cluster analysis is a class of techniques that are used to classify objects or cases into relative groups called clusters. In cluster analysis, there is no prior information about the group or cluster membership for any of the objects. Cluster analysis involves finding groups of objects in a way that the objects in a group will be similar (or related) to one another and different from (or unrelated to) the objects in other groups. That is, a situation where intra-cluster distances are minimized and inter-cluster distances are maximized. Cluster analysis utilizes measures of similarity (and dissimilarity) between any two objects such as Euclidean Distance, Manhattan Distance, Mahalanobis Distance, Minkowski Distance, among others.

When performing cluster analysis, we first partition the set of data into groups based on data similarity and then assign the labels to the groups.

The main advantage of clustering over classification is that it is adaptable to changes and helps single out useful features that distinguish different groups.

Cluster analysis is used in outlier detection applications such as detection of credit card fraud. Cluster analysis can also be used as a data mining function to serve as a tool for gaining insight into the distribution of data to observe the characteristics of each cluster. In clustering, members of the same clusters are closer to each other.

There are three main methods for cluster analysis: two step cluster or K-means cluster or hierarchical cluster.

K-means cluster is a method for quickly clustering large data sets. The aim is to partition observations in K separate clusters that do not overlap. This procedure partitions data into K clusters and K is usually determined by user without looking at the relationships among the clusters. Each cluster is associated with a centroid and each data point is assigned to the cluster with the closest centroid.

Given a set X of n points in a d-dimensional space and an integer k group the points into k clusters $C = \{C_1, C_2, \ldots, C_n\}$ such that

$$\min_{C_1, C_2, \ldots, C_n} \sum_{i=1}^{k} \sum_{x \in C_i} \|x - \mu_i\|^2 .$$

where C_i is the i-th cluster and μ_i is the mean of the points in cluster C_i. The objective is to minimize the within-cluster sum of squares.

Hierarchical cluster is the most common method. It generates a series of models with cluster solutions from 1 (all cases in one cluster) to *n* (each case is an individual cluster). Hierarchical cluster analysis can handle nominal, ordinal and scale data; but not recommended to mix different levels of measurement. This procedure generates a hierarchy of partitions, and the user selects the partition. Hierarchical clustering produces nested clusters that can be visualized as a dendrogram that can either be Agglomerative (bottom up), where initially each point is a cluster and repeatedly combine the two "nearest" clusters into one, or Divisive (top down), where we start with one cluster and recursively split. There are four methods for combining clusters in the Agglomerative approach: single linkage, complete linkage, average linkage, centroid method and Ward's method.

Two-step cluster analysis identifies groupings by running preclustering first and then by running hierarchical methods. It is a combination of the previous two approaches. Two-step clustering can handle scale and ordinal data in the same model, and it automatically selects the number of clusters.

Cluster analysis can be performed in SPSS using Two-Step Cluster, Hierarchical Cluster, or K-Means Cluster, each of them relying on different algorithm to create the clusters.

In SPSS, we go to: Analyze > Classify > select the clustering technique (i.e. Two Step Cluster or K- Means Cluster or **Hierarchical Cluster** > select the variables of interest under Variable(s): box > click Statistics and select Agglomeration schedule then click Continue > click Plots and select Dendogram then click continue > click Method and select the appropriate measurement scale for the data then click Continue then OK. SPSS Output will display the results for Hierarchical Cluster Analysis.

Example 2.9: Using data in Figure 2.1. Assess perform hierarchical cluster analysis.

We go to: Analyze > Classify > select **Hierarchical Cluster** > select the five variables of interest (inc, age, crdbt, odbt, cvalue) under Variable(s): box > click Statistics and select Agglomeration schedule then click Continue > click Plots and select Dendogram then click continue > click Method and select the appropriate measurement scale for the data then click Continue then OK Figure 2.28.

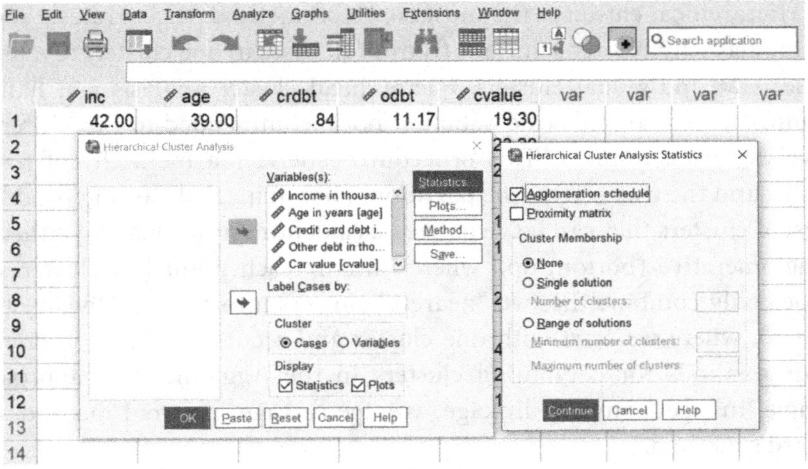

FIGURE 2.28 SPSS Procedure for Hierarchical Cluster Analysis

SPSS Output will display the results for Hierarchical Cluster Analysis Figure 2.29.

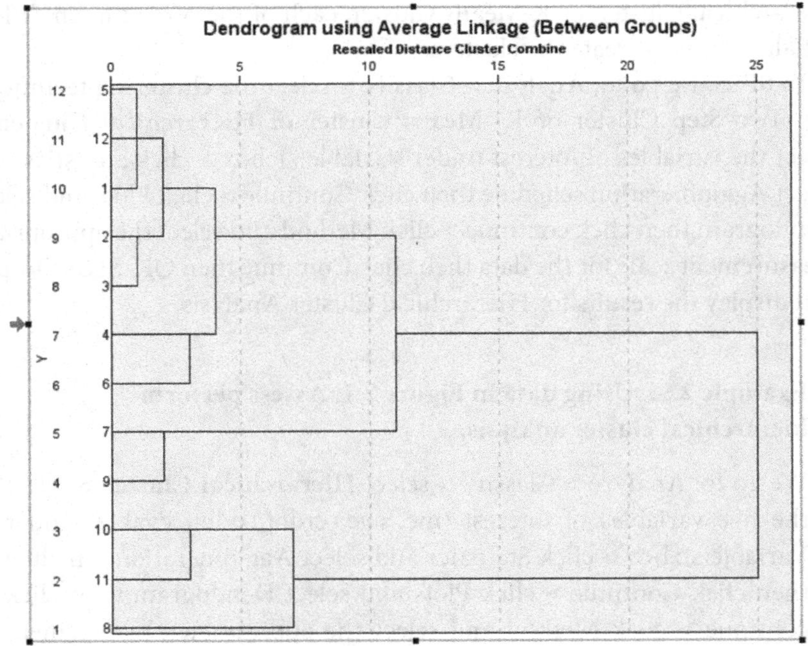

FIGURE 2.29 SPSS Output for the Dendogram

For K-means clustering,

We go to: Analyze > Classify > select K- Means Cluster > select the five variables of interest (inc, age, crdbt, odbt, cvalue) under Variable(s): box then click OK Figure 2.30.

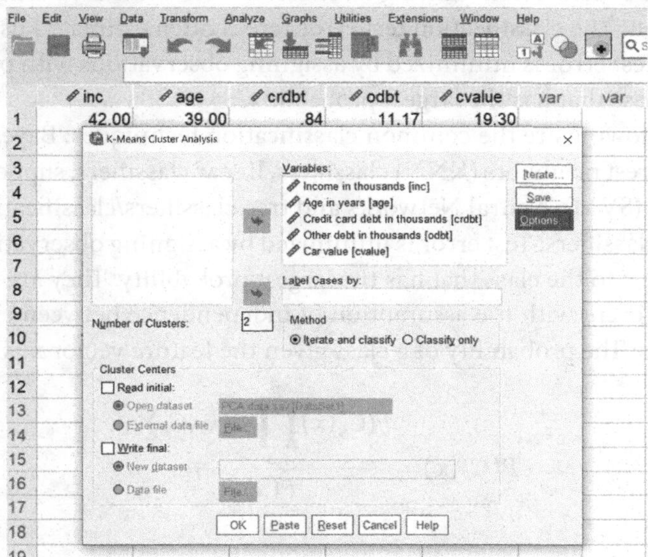

FIGURE 2.30 SPSS procedure for K-Means Clustering

SPSS Output will display the results for K-means cluster analysis Figure 2.31.

Final Cluster Centers

	Cluster 1	Cluster 2
Income in thousands	31.33	89.33
Age in years	43.89	57.67
Credit card debt in thousands	1.08	2.71
Other debt in thousands	2.82	5.04
Car value	15.54	33.13

Number of Cases in each Cluster

Cluster	1	9.000
	2	3.000
Valid		12.000
Missing		.000

FIGURE 2.31 SPSS Output for K-Means Clustering

2.8.2 Classification

Build a model, classifier, etc. to separate data into known groups/classes (supervised learning).

The response variable is not continuous, and we would like to predict labels. The classification training error rate is often estimated using a training data set. The classification test error rate is often estimated using a test data set. Test error is minimized by assigning observations with predictors x to the class that has the largest probability.

The following are the common classification techniques: Bayes classifiers, K nearest neighbors (KNN) classifiers, linear classifiers, support vector machines (SVM), Neural Networks and tree classifiers/classification trees.

Bayes classifiers: Test error is minimized by assigning observations with predictors x to the class that has the largest probability. They are based on Bayes' theorem with the assumption of independence between every pair of features. The probability of a class given the feature vector **x** is given by

$$P(C_k \backslash \mathbf{x}) = \frac{P(C_k \backslash \mathbf{x}) \prod_{i=1}^{p} P(x_i \backslash C_k)}{P(\mathbf{x})}.$$

where:

$P(C_k)$ is the prior probability of class C_k.

$P(x_i \backslash C_k)$ is the likelihood of feature x_i given class C_k.

$P(\mathbf{x})$ is the marginal probability of the feature vector x.

K nearest neighbors (KNN) classifiers: An observation is classified based on the K observations in the training set that are nearest to it. KNN is a non-parametric method for classifying an instance based on the majority label of its k nearest neignbours in the feature space. The Euclidean distance is the most commonly used distance metric given by

$$d(x, y) = \sqrt{\sum_{i=1}^{p} (x_i - y_i)^2}.$$

Linear classifiers: The decision boundary is linear, and we have logistic regression, linear discriminant analysis and quadratic discriminant analysis. For logistic regression, we are interested in predicting two groups with binary labels. For linear discriminant analysis (LDA), the goal is to estimate a decision boundary that gives a classification rule by estimating the posterior probabilities of class membership if an observation is in a particular location, what is the probability it belongs to a particular class. There is also quadratic discriminant analysis (QDA).

Support vector machines (SVM): The goal is to find the hyperplane that "best" separates the two classes (i.e. maximize the margin between the closes points of the classes (support vectors)). The decision function of a binary classification problem is given by

$$f(x) = \mathbf{w}.\mathbf{x} + b.$$

where:

w is the weight vector.
x is the input feature vector.
b is the bias term.

The optimization problem for SVM is:

$$\min_{w,b} \frac{1}{2}\|\mathbf{w}\|^2 \text{ subject to } y_i(\mathbf{w}.\mathbf{x_i} + b) \geq 1 \ \forall i.$$

Neural networks have layers of neurons where each neuron has weights and biases. The output of each neuron is passed via an activation function. For a single layer perception

$$y = \sigma(\mathbf{w}.\mathbf{x} + b).$$

where σ is the activation function (or the sigmoid function).

Classification trees: the goal is to determine which variables are "best" at separating observations into the labelled groups. The predictor space is partitioned into hyper-rectangles, and any observations in the hyper-rectangle would be predicted to have the same label. The next split is chosen to maximize "purity" of hyper-rectangles.

Decision trees are flow-like structure where the internal node represents a feature or an attribute and each branch represents a decision rule and each leaf node represents an outcome or class label. The decision tree splits the data recursively based on the feature that gives in the highest information gain or Gini impurity reduction.

The information gain is given by

$$IG(T,a) = H(T) - \sum_{v \in values(a)} \frac{|T_v|}{|T|} H(T_v).$$

where

$H(T)$ is the entropy of the entire set T.
T_v is the subset of T for which attribute a has value v.

The Gini Impurity is given by

$$G(T)=1-\sum_{i=1}^{c} p_i^2.$$

where

p_i is the probability of class i in set $\boldsymbol{T}$.
c is the total number of classes.

To perform statistical learning in SPSS, we go to: Analyze> Classify > select Naïve Bayes or Tree or Discriminant or Nearest Neighbor.

Example 2.10: Using data in Figure 2.12, perform naïve Bayes classification.

We go to: Analyze> Classify > select Naïve Bayes > select level of education as the Dependent variable > select the other numeric variables under Variables to Exclude Figure 2.32.

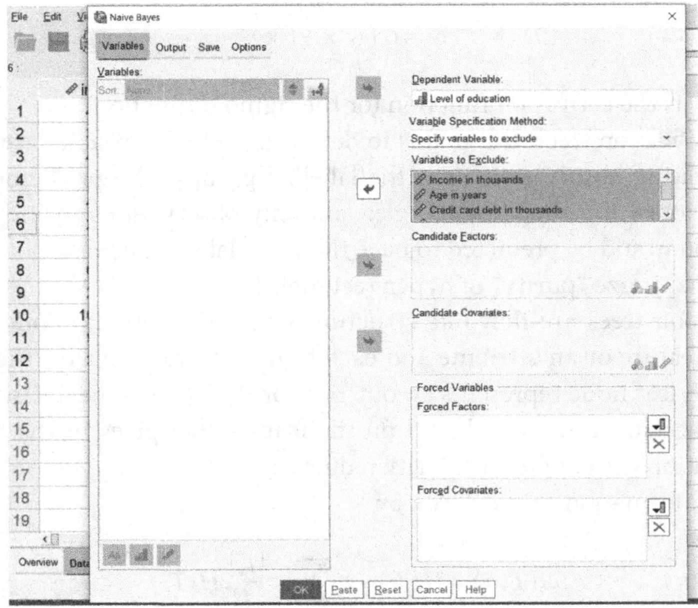

FIGURE 2.32 SPSS Procedure for Naive Bayes

SPSS Output will display the results of Naive Bayes Classification.

To perform other statistical learning in SPSS, we use IBM SPSS Modeler for Predictive Analytics, Machine Learning and AI.

2.9 CORRESPONDENCE ANALYSIS

Correspondence analysis takes a different sort of approach to figuring out where data changes the most. Correspondence analysis looks for the directions where the data are "most surprising" from a chi-squared test perspective.

Correspondence analysis is appropriate when attempting to determine the proximal relationships among two or more categorical variables. Using correspondence analysis with categorical variables is analogous to using correlation analysis and principal components analysis for continuous variables. They provide insights to the relationships among variables and the dimensions or eigenvectors underlying them.

From correspondence analysis, the multidimensional map is produced. The correspondence map allows to visualize the relationships among categories spatially on dimensional axes i.e. which categories are close to other categories on empirically derived dimensions.

Correspondence analysis is a method for visualizing the rows and columns of a table of non-negative data as points in a map, with a specific spatial interpretation. Data are usually counted in a cross-tabulation. For cross-tabulations, the method explains the association between the rows and columns of the table as measured by the Pearson chi-square statistic. The method has several similarities to principal component analysis, in that it situates the rows or the columns in a high-dimensional space and then finds a best-fitting subspace, usually a plane, in which to approximate the points.

A correspondence table is any rectangular two-way array of non-negative quantities that indicates the strength of association between the row entry and the column entry of the table. The most common example of a correspondence table is a contingency table, in which row and column entries refer to the categories of two categorical variables, and the quantities in the cells of the table are frequencies.

Initially, we have a contingency table **N** with I rows and J columns. The element n_{ij} represents the frequency count for the i-th column. We then compute the Correspondence matrix by converting the contingency table into correspondence matrix P as

$$\mathbf{P} = \frac{1}{n}\mathbf{N}.$$

where $n = \sum_{i=1}^{I} \sum_{j=1}^{J} n_{ij}$.

We then calculate the row and marginal totals (proportions) as $\mathbf{r} = \mathbf{P}1_j$ and $\mathbf{c} = \mathbf{P}^T 1_I$, where 1_j and 1_I are vectors of ones of length J and I, respectively.

We then standardize the Correspondence Matrix by calculating the standardized residuals (the matrix of deviations from independence) as

$$\mathbf{S} = \mathbf{D}_r^{-\frac{1}{2}} (\mathbf{P} - \mathbf{rc}^T) \mathbf{D}_c^{-\frac{1}{2}}.$$

where $\mathbf{D}_r$ and $\mathbf{D}_c$ are diagonal matrices of the rows and column marginal totals, respectively.

We then perform Singular Value Decomposition (SVD) on the standardized residual matrix $\mathbf{S}$:

$$\mathbf{S} = \mathbf{U} \sum \mathbf{V}^T.$$

where:

$\mathbf{U}$ is an $I \times I$ orthogonal matrix (left singular vectors).
Σ is an $I \times J$ diagonal matrix of singular values.
$\mathbf{V}$ is an $J \times J$ orthogonal matrix (right singular vectors).

Finally, we calculate the principal coordinates for the rows and columns and obtain

$$\mathbf{F} = \mathbf{D}_r^{-\frac{1}{2}} \mathbf{U} \sum \quad \text{and} \quad \mathbf{G} = \mathbf{D}_c^{-\frac{1}{2}} \mathbf{V} \sum.$$

To perform Correspondence Analysis in SPSS, we go to: Analyze > Data Reduction > Correspondence Analysis > select the categorical variable of interest under Row and Define the Row Range by putting the codes (smallest to the highest) for the categorical variable then click Continue > select the other categorical variable of interest under Column and Define Column Ranges by putting the codes (smallest to the highest) for the categorical variable then click Continue > click Statistics and select Correspondence table, Overview of row points, Overview of column points, Row profiles, Column profiles as well as Confidence Statistics for Row points and Column points then click the Continue > click Plots and select Biplot, Row points, Column points, Transformed row categories and

Transformed column categories then click the Continue button, then click the OK button. SPSS Output will display the results for Correspondence Analysis.

*Correspondence Data.sav [DataSet3] - IBM SPSS Statistics Data Editor

File Edit View Data Transform Analyze Graphs Utilities Extensions Window

	edcat	jobcat
1	College degree	Managerial and Professional
2	Some college	Sales and Office
3	High school degree	Sales and Office
4	Did not complete high school	Managerial and Professional
5	Some college	Service
6	High school degree	Precision Production, Craft, Rep...
7	High school degree	Sales and Office
8	Did not complete high school	Managerial and Professional
9	Some college	Agricultural and Natural Resources
10	Did not complete high school	Managerial and Professional
11	College degree	Agricultural and Natural Resources
12	Some college	Sales and Office

FIGURE 2.33 Data for Correspondence Analysis

Example 2.11: Consider the correspondence data (Figure 2.33). Perform correspondence analysis.

We go to: Analyze > Data Reduction > Correspondence Analysis > select level of education under Row and click Define Range and put the codes (minimum value = 1 to maximum value = 5) for level of education then click Update then Continue > select the job category under Column and click Define Range then put the codes (minimum value = 1 to maximum value = 6) for the job category then click Update then Continue > click Statistics and select Correspondence table, Overview of row points, Overview of column points, Row profiles, Column profiles as well as Confidence Statistics for Row points and Column points then click the Continue > click Plots and select Biplot, Row points, Column points, Transformed row categories and Transformed column categories then click the Continue button, then click the OK button (Figure 2.34).

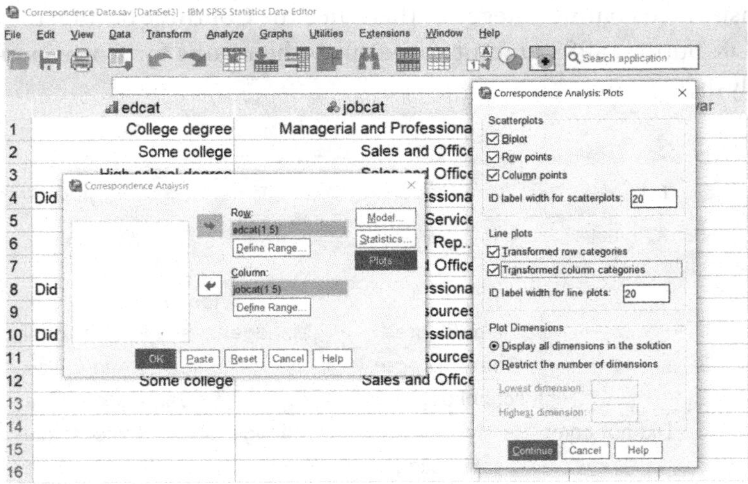

FIGURE 2.34 SPSS Procedure for Correspondence Analysis

SPSS Output will display the results for Correspondence Analysis in terms of correspondence table (Figure 2.35), summary (Figure 2.36) plus row and column points (Figure 3.37).

Correspondence Table

	Job category					
Level of education	Managerial and Professional	Sales and Office	Service	Agricultural and Natural Resources	Precision Production, Craft, Repair	Active Margin
Did not complete high school	3	0	0	0	0	3
High school degree	0	2	0	0	1	3
Some college	0	2	1	1	0	4
College degree	1	0	0	1	0	2
Post-undergraduate degree	0	0	0	0	0	0
Active Margin	4	4	1	2	1	12

Row Profiles

	Job category					
Level of education	Managerial and Professional	Sales and Office	Service	Agricultural and Natural Resources	Precision Production, Craft, Repair	Active Margin
Did not complete high school	1.000	.000	.000	.000	.000	1.000
High school degree	.000	.667	.000	.000	.333	1.000
Some college	.000	.500	.250	.250	.000	1.000
College degree	.500	.000	.000	.500	.000	1.000
Post-undergraduate degree	.000	.000	.000	.000	.000	.000
Mass	.333	.333	.083	.167	.083	

FIGURE 2.35 SPSS Output for Correspondence Table

Column Profiles

Level of education	Job category: Managerial and Professional	Sales and Office	Service	Agricultural and Natural Resources	Precision Production, Craft, Repair	Mass
Did not complete high school	.750	.000	.000	.000	.000	.250
High school degree	.000	.500	.000	.000	1.000	.250
Some college	.000	.500	1.000	.500	.000	.333
College degree	.250	.000	.000	.500	.000	.167
Post-undergraduate degree	.000	.000	.000	.000	.000	.000
Active Margin	1.000	1.000	1.000	1.000	1.000	

Summary

Dimension	Singular Value	Inertia	Chi Square	Sig.	Proportion of Inertia: Accounted for	Proportion of Inertia: Cumulative	Confidence Singular Value: Standard Deviation	Confidence Singular Value: Correlation 2
1	.937	.879			.620	.620	.034	-.097
2	.638	.407			.287	.907	.131	
3	.362	.131			.093	1.000		
Total		1.417	17.000	.386[a]	1.000	1.000		

a. 16 degrees of freedom

FIGURE 2.36 SPSS Output for Summary of Correspondence Analysis

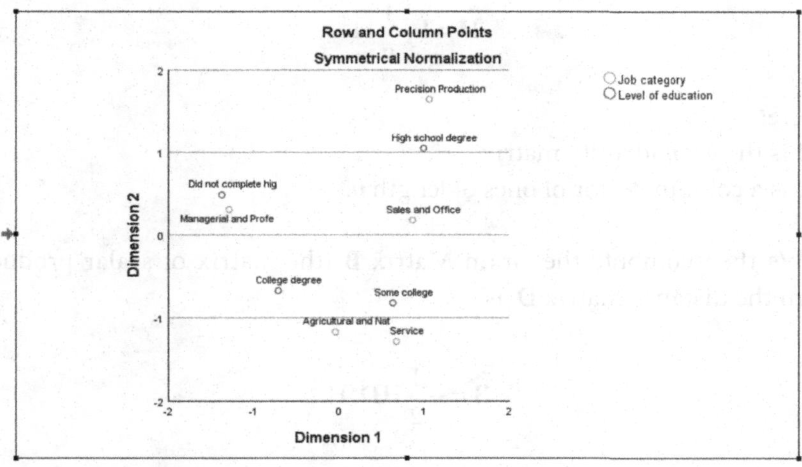

FIGURE 2.37 SPSS Output for Row and Column Points

2.10 MULTIDIMENSIONAL SCALING

In some cases, we only have a distance matrix to work with, but we want to visualize the data so that points, which are close to each other in the distance matrix, appear close to each other in a figure. One way to do this is with multidimensional scaling.

Multidimensional scaling (MDS) is a technique that creates a map displaying the relative positions of several objects, given only a table of the distances between them. The map may consist of one, two, three or even more dimensions. The program calculates either the metric or the nonmetric solution. The table of distances is known as the proximity matrix. It arises either directly from experiments or indirectly as a correlation matrix.

MDS is a means of coordinating (i.e. creating categories and clines) and visualizing data by taking potentially complex information and arranging it into a set of points in n-dimensional space i.e. 1-dimensional space: a line e.g. a number line; two-dimensional space: a plane/a surface; three-dimensional space: a volume e.g. a cube; and four- or higher dimensional space: no physical analog.

We first compute the distance matric **D** for n objects where d_{ij} is the distance between objects i and j.

We then define the centering matrix **H** as

$$\mathbf{H} = \mathbf{I} - \frac{1}{n}11^T.$$

where:

I is the $n \times n$ identity matrix.

1 is a column vector of ones of length n.

We then compute the Gram Matrix **B** (the matrix of scalar product) from the distance matrix **D** as

$$\mathbf{B} = -\frac{1}{2}\mathbf{H}\mathbf{D}^2\mathbf{H}.$$

where $\mathbf{D}^2$ is the element-wise square of the distance matrix.

We then perform eigenvalue decomposition on the Gram matrix B:

$$\mathbf{B} = \mathbf{V}\boldsymbol{\Lambda}\mathbf{V}^T.$$

where:

V is the matrix of eigenvectors.

Λ is the diagonal matrix of eigenvalues.

Finally, we compute the coordinates of the objects in the lower dimensional space as

$$\mathbf{X}_m = \mathbf{V}_m \Lambda_m^{\frac{1}{2}}$$

where:

$\mathbf{V}_m$ is the diagonal matrix of the top m eigenvalues.
Λ_m is the matrix of the corresponding eigen vectors.

To perform Multidimensional Scaling Analysis in SPSS we go to: Analyze > Scale > Multidimensional Scaling (ALSCAL) > Select all the variables you want to include in the analysis > click Model and select Ordinal Level of Measurement, Matrix Conditionality and Euclidean distance Scaling Model then click Continue > click Options and select Group plots then click Continue then OK. SPSS Output will display the results for the Multidimensional Scaling.

Example 2.12: Using multidimensional data. Perform multidimensional scaling analysis. We first estimate the mean comparisons between different products.

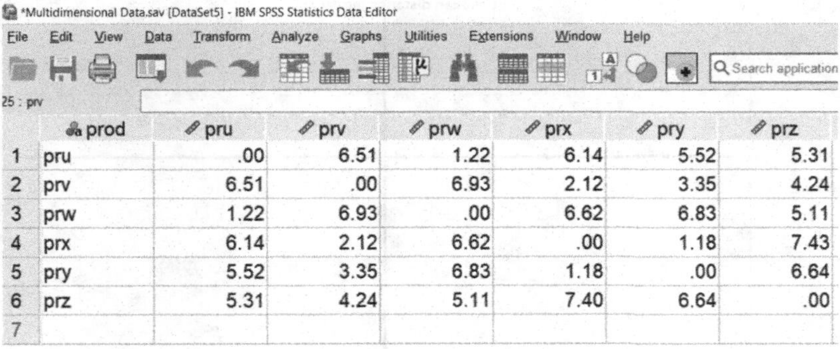

*Multidimensional Data.sav [DataSet5] - IBM SPSS Statistics Data Editor

File Edit View Data Transform Analyze Graphs Utilities Extensions Window Help

25 : prv

	prod	pru	prv	prw	prx	pry	prz
1	pru	.00	6.51	1.22	6.14	5.52	5.31
2	prv	6.51	.00	6.93	2.12	3.35	4.24
3	prw	1.22	6.93	.00	6.62	6.83	5.11
4	prx	6.14	2.12	6.62	.00	1.18	7.43
5	pry	5.52	3.35	6.83	1.18	.00	6.64
6	prz	5.31	4.24	5.11	7.40	6.64	.00
7							

FIGURE 2.38 Multidimensional Scaling Data

We go to: Analyze > Scale > Multidimensional Scaling (ALSCAL) > Select all numeric variables (pru to prz) > click Model and select Interval Level of Measurement, Matrix Conditionality and Euclidean distance Scaling Model then click Continue > click Options and select Group plots then click Continue then OK.

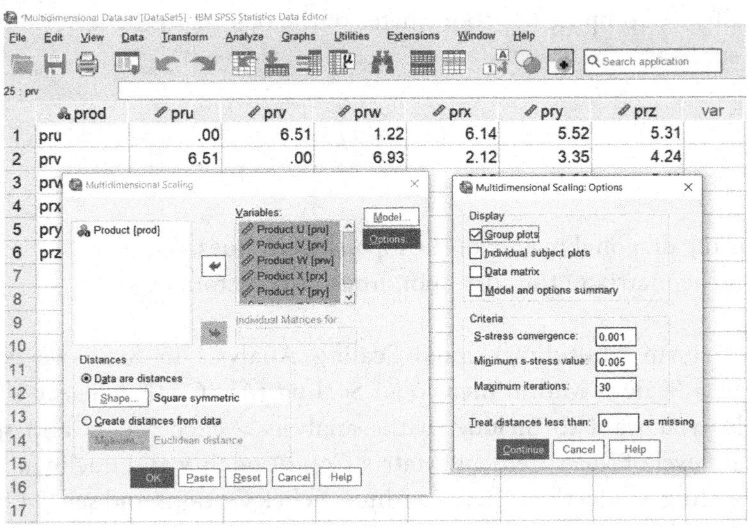

FIGURE 2.39 SPSS Procedure for Multidimensional Scaling Analysis

SPSS Output will display the results for the Multidimensional Scaling – perceptual map (Figure 2.40).

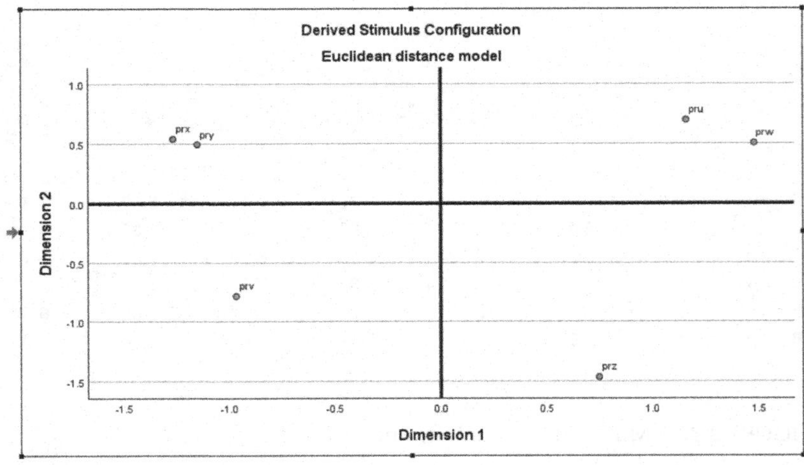

FIGURE 2.40 Multidimensional Scaling Analysis SPSS Output displaying Perceptual Map

The two-dimensional graph where six products are grouped into four segments. Prv has no similarity with other products.

2.11 CANONICAL CORRELATION ANALYSIS

Canonical correlation analysis is used to identify and measure the associations among two sets of variables. Canonical correlation analysis determines a set of canonical variates, orthogonal linear combinations of the variables within each set that best explain the variability both within and between sets. The canonical correlation is a multivariate analysis of correlation. Canonical is the statistical term for analyzing latent variables (which are not directly observed) that represent multiple variables (which are directly observed). Canonical Correlation analysis is the analysis of multiple-X multiple-Y correlation. The canonical correlation coefficient measures the strength of association between two canonical variates. A canonical variate is the weighted sum of the variables in the analysis.

Assumptions

- Canonical correlation analysis requires the multivariate normal and homogeneity of variance assumption.
- Canonical correlation analysis assumes a linear relationship between the canonical variates and each set of variables.
- Canonical correlation analysis requires a large sample size.

Initially, we have two sets of variables **X** and **Y** with n observations each where **X** is an $n \times p$ matrix and **Y** is an $n \times q$ matrix.

We then compute the covariance matrices for **X** and **Y** as

$$\mathbf{S}_{XX} = \frac{1}{n-1}\mathbf{X}^T\mathbf{X}.$$

$$\mathbf{S}_{YY} = \frac{1}{n-1}\mathbf{Y}^T\mathbf{Y}.$$

$$\mathbf{S}_{XY} = \frac{1}{n-1}\mathbf{X}^T\mathbf{Y}.$$

We then compute canonical correlation matrices by estimating vectors **a** and **b** that maximize the correlation between **Xa** and **Yb**.

For **X**: $\mathbf{S}_{XX}^{-1}\mathbf{S}_{YX}\mathbf{S}_{XX}^{-1}\mathbf{S}_{YX}\mathbf{a} = \rho^2\mathbf{a}$ and for **Y**: $\mathbf{S}_{YY}^{-1}\mathbf{S}_{YX}\mathbf{S}_{XX}^{-1}\mathbf{S}_{XY}\mathbf{a} = \rho^2\mathbf{b}$, where ρ is the canonical correlation while **a** and **b** are the canonical vectors.

We then solve the generalized eigenvalue problem by finding the eigenvalues ρ^2 and the corresponding vectors **a** and **b**.

We finally form the canonical variables (canonical scores) as $\mathbf{U} = \mathbf{XA}$ and $\mathbf{V} = \mathbf{YB}$, where **A** and **B** are matrices containing canonical vectors **a** and **b** as columns.

To perform Canonical Correlation Analysis in SPSS, we go to: Analyze > Correlate > Canonical correlation > select the set of independent variables as Set 1 and set of dependent variables as Set 2 then click OK. SPSS Output will display the results of the Canonical Correlation Analysis.

Example 2.13: Consider the multivariate data with five scale/numeric variables. We consider three independent variables (crdbt, odbt, cvalue) as one canonical variable and two dependent variables (inc, age) as one canonical variable.

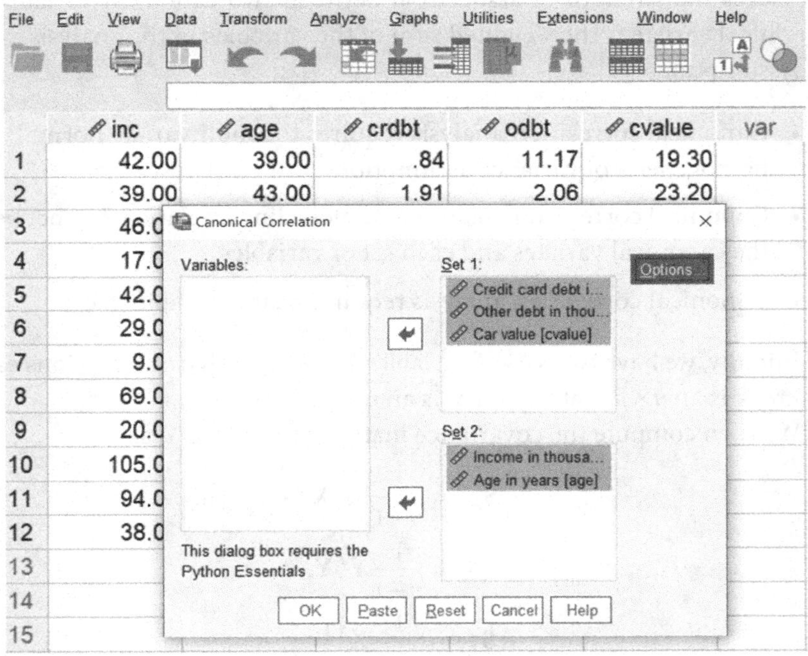

FIGURE 2.41 Multivariate Data

In SPSS, we go to: Analyze > Correlate > Canonical correlation > select the set of independent variables (crdbt, odbt, cvalue) as Set 1 and set of dependent variables (inc, age) as Set 2 then click OK. SPSS Output will display the results of the Canonical Correlation Analysis as shown in Figure 2.42. The findings shows that the two correlations are statistically significant. The findings are also presented for the canonical loadings for set 1 and set 2 (Figure 2.43).

Canonical Correlations Settings

	Values
Set 1 Variables	crdbt odbt cvalue
Set 2 Variables	inc age
Centered Dataset	None
Scoring Syntax	None
Correlations Used for Scoring	2

Canonical Correlations

	Correlation	Eigenvalue	Wilks Statistic	F	Num D.F	Denom D.F.	Sig.
1	.918	5.363	.148	3.739	6.000	14.000	.020
2	.246	.064	.940	.	.	.	.

H0 for Wilks test is that the correlations in the current and following rows are zero

Set 1 Canonical Loadings

Variable	1	2
crdbt	-.814	.319
odbt	-.574	-.797
cvalue	-.997	.021

Set 2 Canonical Loadings

Variable	1	2
inc	-.997	.079
age	-.051	.999

Set 1 Cross Loadings

Variable	1	2
crdbt	-.747	.078
odbt	-.527	-.196
cvalue	-.916	.005

Set 2 Cross Loadings

Variable	1	2
inc	-.915	.019
age	-.047	.246

Proportion of Variance Explained

Canonical Variable	Set 1 by Self	Set 1 by Set 2	Set 2 by Self	Set 2 by Set 1
1	.662	.558	.498	.420
2	.246	.015	.502	.030

FIGURE 2.42 SPSS Procedure for Canonical Correlation Analysis

2.12 PRACTICE EXERCISE

1. Classify the different techniques for analyzing multivariate data.
2. Using appropriate SPSS data set, generate and interpret appropriate graphs for visualizing multivariate data.
3. Using appropriate SPSS data set, assess multivariate normality using
   a. Appropriate graphical technique.
   b. Appropriate statistical test.
4. Using appropriate SPSS data set, perform and interpret the following analysis:
   a. Different inferences for multivariate means.
   b. Different data reduction techniques.
   c. Different statistical learning techniques.
5. Using appropriate SPSS data set, perform and interpret the following analysis:
   a. Correspondence analysis.
   b. Multidimensional scaling analysis.
   c. Canonical correlation analysis.

CHAPTER 3

Survival Analysis Using SPSS

3.1 INTRODUCTION

Survival analysis is concerned with analyzing time to an event of interest data (survival data). A time to event variable reflects the time until a participant has an event of interest (e.g. time to heart attack, time to cancer remission, time to death). Times to event are always positive, and their distributions are often skewed.

Survival data can be described and modeled in terms of two related probabilities: survival and hazard. It is expressed as

$$S(t) = P(T > t).$$

where T is the random variable representing the time to event.

The survival probability (survivor function) $S(t)$ is the probability that an individual survives from the time origin to a specified future time t.

The hazard is usually denoted by $h(t)$ or $\lambda(t)$ and is the probability that an individual who is under observation at a time t has an event at that time. It represents the instantaneous event rate for an individual who has already survived to time t. The hazard function is given by

$$h(t) = \lim_{\Delta t \to 0} \frac{P\left(t \le T < t + \Delta t \middle| T \ge t\right)}{\Delta t}.$$

$$h(t) = -\frac{d}{dt}\left[logS(t)\right].$$

DOI: 10.1201/9781003386636-3

The cumulative hazard function $H(t)$ is the integral of the hazard function over time and is given by

$$H(t) = \int_0^t h(u)du.$$

$$H(t) = -log\left[S(t)\right].$$

3.1.1 Censoring

There are several different types of censoring. The most common is called right censoring and occurs when a participant does not have the event of interest during the study, and thus their last observed follow-up time is less than their time to event. This can occur when a participant drops out before the study ends or when a participant is event free at the end of the observation period.

3.1.2 The Survival Function

In survival analysis, we use information on event status and follow up time to estimate a survival function.

Survival time is a common response for studying the effects of treatments on lifetime. The properties of survival times are usually characterized by (1) survival function (also known as reliability function or cumulative survival rate) and (2) hazard function (also known as failure rate function). The survival time often is censored. It may be left-censored, right-censored or both. A hazard function of survival time T is the conditional failure rate defined as the probability of failure during a small time interval given the individual has survived. Survival analysis usually studies the survival time based on some treatment effects and the covariates.

There are four different ways of studying survival time: life tables, Kaplan-Meier, Cox regression model and Cox regression with time-dependent covariates.

3.2 LIFE TABLES

The first step of constructing a lift timetable is to subdivide the period of observation into smaller time intervals. For each small time interval, the subjects, which are observed at least that long, are used to calculate the probability of a terminal event occurring in that interval. The probabilities estimated from each of the intervals are then used to estimate the overall probability of the event occurring at different time points.

In the life table, we have n_i which is the number of individuals who are at risk of experiencing the event before time t_i. We have d_i, which represents the number of individuals who experience the event (e.g. death, failure) at time t_i. We have c_i representing the number of individuals who are censored (i.e. withdrawn from the study or lost to follow-up) between t_i and t_{i+1}. We have $S(t)$ which represents the probability that an individual survives from the time origin to a specified time t.

The probability of a given event occurring at time t_i is given by

$$q_i = \frac{d_i}{n_i}.$$

The probability of surviving through the interval time t_i and $t_i + 1$ is given by

$$p_i = 1 - q_i = 1 - \frac{d_i}{n_i}.$$

The cumulative survival probability up to time t_i is given by

$$S(t_i) = \prod_{j=1}^{i} p_j = \prod_{j=1}^{i} \left(1 - \frac{d_j}{n_j}\right).$$

The adjusted number at risk to account for individuals censored during the interval is given by

$$\tilde{n}_i = n_i - \frac{c_i}{2}.$$

A typical life table will have the time interval (t_i), number at risk (n_i), number of events (d_i), number censored (c_i), probability of events q_i, probability of survival (p_i) and cumulative survival probability ($S(t_i)$).

To generate Life Tables in SPSS, we go to: Analyze > Survival > Life Tables > put the time variable into Time field and specify the time intervals > put the status variable under the Status field and Define the Event > click Option and select Life Tables then click OK. SPSS Output will generate results for Life Table.

Example 3.1: Consider the survival data (Figure 3.1). Generate a life table for tenure.

Survival Data.sav [DataSet6] - IBM SPSS Statistics Data Editor

	marital	age	region	gender	ed	tenure	default	var
1	Married	39	Zone 2	Female	17	3	Yes	
2	Unmarried	43	Zone 4	Male	16	39	Yes	
3	Married	49	Zone 1	Male	14	65	No	
4	Unmarried	35	Zone 5	Male	9	25	No	
5	Married	32	Zone 5	Male	16	42	No	
6	Unmarried	21	Zone 4	Female	14	2	Yes	
7	Unmarried	79	Zone 3	Female	12	72	No	
8	Unmarried	64	Zone 1	Male	11	66	No	
9	Married	62	Zone 3	Male	16	72	No	
10	Married	61	Zone 1	Male	9	41	No	
11	Married	48	Zone 2	Female	17	38	No	
12	Married	35	Zone 5	Female	16	7	No	
13	Married	25	Zone 5	Female	18	39	Yes	
14	Married	61	Zone 4	Female	16	57	No	

FIGURE 3.1 Survival Data

We go to: Analyze > Survival > Life Tables > put the time variable (tenure) into Time field and specify the time intervals (put 60 as maximum and 3 as the number of intervals) > put the status variable (Default) under the Status field and Define the Event (set 1) > click Option and select Life Tables then click OK (Figure 3.2).

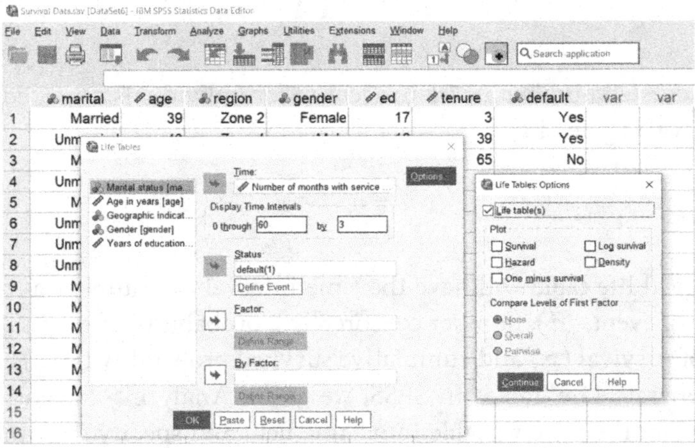

FIGURE 3.2 SPSS Procedure for Generating Life-Table

SPSS Output will generate results for Life Table (Figure 3.3).

Life Table[a]

Interval Start Time	Number Entering Interval	Number Withdrawing during Interval	Number Exposed to Risk	Number of Terminal Events	Proportion Terminating	Proportion Surviving	Cumulative Proportion Surviving at End of Interval	Std. Error of Cumulative Proportion Surviving at End of Interval	Probability Density	Std. Error of Probability Density	Hazard Rate	Std. Error of Hazard Rate
0	14	0	14.000	1	.07	.93	.93	.07	.024	.023	.02	.02
3	13	0	13.000	1	.08	.92	.86	.09	.024	.023	.03	.03
6	12	1	11.500	0	.00	1.00	.86	.09	.000	.000	.00	.00
9	11	0	11.000	0	.00	1.00	.86	.09	.000	.000	.00	.00
12	11	0	11.000	0	.00	1.00	.86	.09	.000	.000	.00	.00
15	11	0	11.000	0	.00	1.00	.86	.09	.000	.000	.00	.00
18	11	0	11.000	0	.00	1.00	.86	.09	.000	.000	.00	.00
21	11	0	11.000	0	.00	1.00	.86	.09	.000	.000	.00	.00
24	11	1	10.500	0	.00	1.00	.86	.09	.000	.000	.00	.00
27	10	0	10.000	0	.00	1.00	.86	.09	.000	.000	.00	.00
30	10	0	10.000	0	.00	1.00	.86	.09	.000	.000	.00	.00
33	10	0	10.000	0	.00	1.00	.86	.09	.000	.000	.00	.00
36	10	1	9.500	0	.00	1.00	.86	.09	.000	.000	.00	.00
39	9	1	8.500	2	.24	.76	.66	.14	.067	.042	.09	.06
42	6	1	5.500	0	.00	1.00	.66	.14	.000	.000	.00	.00
45	5	0	5.000	0	.00	1.00	.66	.14	.000	.000	.00	.00
48	5	0	5.000	0	.00	1.00	.66	.14	.000	.000	.00	.00
51	5	0	5.000	0	.00	1.00	.66	.14	.000	.000	.00	.00
54	5	0	5.000	0	.00	1.00	.66	.14	.000	.000	.00	.00
57	5	1	4.500	0	.00	1.00	.66	.14	.000	.000	.00	.00
60	4	4	2.000	0	.00	1.00	.66	.14	.000	.000	.00	.00

a. The median survival time is 60.00

FIGURE 3.3 SPSS Output Displaying Life-Table

3.3 KAPLAN-MEIER

An issue with the life table approach is that the survival probabilities can change depending on how the intervals are organized, particularly with small samples. The Kaplan-Meier (KM) approach (product-limit approach) addresses this issue by re-estimating the survival probability each time an event occurs.

Kaplan-Meier method is a nonparametric technique for estimating the survival rates with the presence of censored cases. The basic idea is to first compute the conditional probabilities at each time point when an event occurs and then compute the product limit of those probabilities to estimate the survival rate at each point in time. It is also called Product-Limit method. This technique is often used for comparing the effects of treatments on the survival time. The survival probability can be estimated nonparametrically from observed survival times, both censored and uncensored, using the KM (or product-limit) method.

$$\hat{S}(t)=\prod_{t_i \leq t}\left(1-\frac{d_i}{n_i}\right).$$

where:

t_i is the ordered event times.
d_i is the number of events at time t_i.
n_i is the number of individuals at risk just before time t_i.

To compare two survival curves for different groups, we use log rank test. It tests the null hypothesis that there is no difference in survival between the groups. The log-rank tests statistics compares the observed and expected events across all the time points and its given by

$$\chi^2=\sum_i \frac{\left(O_{ij}-E_{ij}\right)^2}{E_{ij}}.$$

with $df=k-1$, where k is the number of groups, O_{ij} is the number of events observe in group j at time t_i and E_{ij} is the number for events expected in group j at time t_i and is computed as

$$E_{ij}=d_i\frac{n_{i,j}}{n_i}.$$

where:

d_i is the total number of events at time t_i.
$n_{i,j}$ is the number of individuals at risk in group j just before time t_i.
n_i is the total number for individuals at risk just before time t_i.

For a Kaplan-Meier survival analysis, we need at least four variables: ID (case identifier/participant), survival time, event status (1 – event and 0 – censored) and the between-subjects factor (the two groups that we are comparing).

To perform Kaplan-Meier analysis in SPSS, we go to: Analyze > Survival > Kaplan-Meier > Transfer the survival time variable into the Time: box > Transfer the event status variable into the Status: box > click Define Event and put 1 under Single value then click Continue > If you want to compare factors/groups, then select categorical variable for comparison and drag it to factor box > Click Options and select Survival table(s) and Mean and median survival in the Statistics area and select the Survival in the Plots area then click Continue > click Compare Factor and select the Log rank and Pooled over strata then click Continue then OK. SPSS Output will give Kaplan-Meier results.

Example 3.2: Generate the Kaplan-Meier survival curve for default by gender using data in Figure 3.1.

We must define the event of interest, which must have two outcomes i.e. binary plus the follow-up variable.

We go to: Analyze > Survival > Kaplan-Meier > Transfer the survival time variable (tenure) into the Time: box > Transfer the event status variable (default) into the Status: box > click Define Event and put 1 under Single value then click Continue > If you want to compare factors/groups, then select gender and drag it to factor box > Click Options and select Survival table(s) and Mean and median survival in the Statistics area and select the Survival in the Plots area then click Continue > click Compare Factor and select the Log rank and Pooled over strata then click Continue then OK (Figure 3.4).

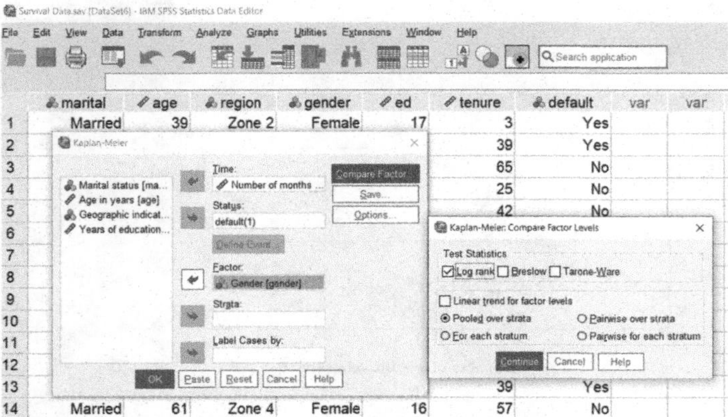

FIGURE 3.4 SPSS Procedure for Kaplan-Meier

SPSS Output will give Kaplan-Meier results for survival table (Figure 3.5) and overall comparison with survival curves (Figure 3.6).

Survival Table

Gender		Time	Status	Cumulative Proportion Surviving at the Time: Estimate	Std. Error	N of Cumulative Events	N of Remaining Cases
Male	1	25.000	No	.	.	0	6
	2	39.000	Yes	.833	.152	1	5
	3	41.000	No	.	.	1	4
	4	42.000	No	.	.	1	3
	5	65.000	No	.	.	1	2
	6	66.000	No	.	.	1	1
	7	72.000	No	.	.	1	0
Female	1	2.000	Yes	.857	.132	1	6
	2	3.000	Yes	.714	.171	2	5
	3	7.000	No	.	.	2	4
	4	38.000	No	.	.	2	3
	5	39.000	Yes	.476	.225	3	2
	6	57.000	No	.	.	3	1
	7	72.000	No	.	.	3	0

Means and Medians for Survival Time

Gender	Mean[a] Estimate	Std. Error	95% Confidence Interval Lower Bound	Upper Bound	Median Estimate	Std. Error	95% Confidence Interval Lower Bound	Upper Bound
Male	66.500	5.021	56.659	76.341	.	.	.	.
Female	44.286	11.872	21.017	67.555	39.000	.	.	.
Overall	55.786	7.012	42.042	69.529	.	.	.	.

a. Estimation is limited to the largest survival time if it is censored.

FIGURE 3.5 SPSS Output for Kaplan-Meier

Overall Comparisons

	Chi-Square	df	Sig.
Log Rank (Mantel-Cox)	2.121	1	.145

Test of equality of survival distributions for the different levels of Gender.

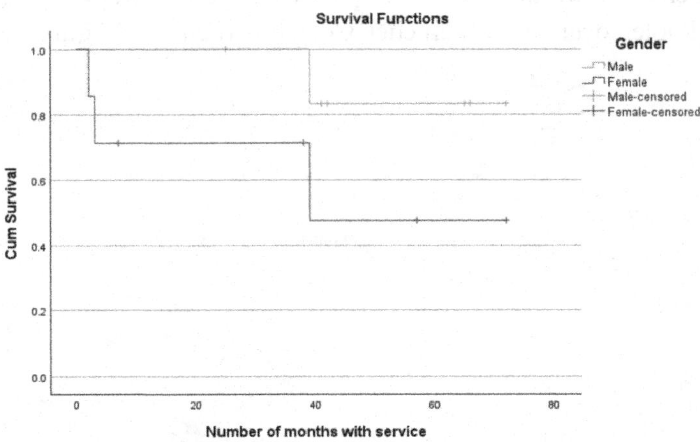

FIGURE 3.6 SPSS Output for Kaplan-Meier Survival Curves

The difference in default is not statistically significant between the two genders.

3.4 COX PROPORTIONAL HAZARD REGRESSION MODEL

Cox regression model is a common technique used for comparing the survival time among treatment levels and taking into account the covariate effects with the presence of censored cases. This is also known as a proportional hazard model. Proportional hazard model assumes that the covariate effects on a hazard function are the same for different factor levels for all time points. That is, the ratio of the hazard functions for two individuals with values of covariate vectors x_1 and x_2 does not vary with time *t*.

The cox regression model is given by

$$h(t\,|\,\mathbf{x}) = h_0(t)\,\boldsymbol{exp}\left(\beta_1 X_1 + \beta_2 X_2 + \ldots + \beta_p X_p\right).$$

where:

$h(t)$ is the expected hazard at time *t*.

$h_0(t)$ is the baseline hazard and represents the hazard when all of the predictors (or independent variables) X_1, X_2, X_p are equal to zero.

$\mathbf{x}$ is the vector of covariates and $\boldsymbol{\beta}$ is a vector of coefficients.

The parameters are estimated by maximizing the partial likelihood function.

Cox proportional hazards regression is used to relate several risk factors or exposures, considered simultaneously, to survival time. In a Cox proportional hazards regression model, the measure of effect is the hazard rate, which is the risk of failure (i.e. the risk or probability of suffering the event of interest), given that the participant has survived up to a specific time.

Assumptions for Cox proportional hazards regression model:

1. Independence of survival times between distinct individuals in the sample.
2. A multiplicative relationship between the predictors and the hazard.
3. A constant hazard ratio over time.

To generate Cox regression model in SPSS, we go to: Analyze > Survival > Cox Regression > Transfer the survival time variable into the Time: box > Transfer the event status variable into the Status: box > click Define Event and put 1 under Single value then click Continue > select the variables (categorical or numeric) to adjust for the covariates and drag them to Covariates box. If categorical, click Categorical and move to Categorical Covariates box > click Plots and select Survival ten Continue > click Options and CI of exp(B) then click Continue then OK. SPSS Output will give the results of Cox regression model.

Example 3.3: Consider the survival data. Perform Cox regression analysis.

We go to: Analyze > Survival > Cox Regression > Transfer the survival time variable (tenure) into the Time: box > Transfer the event status variable (default) into the Status: box > click Define Event and put 1 under Single value then click Continue > select the variables (gender) to adjust for the covariates and drag them to Covariates box. If categorical click Categorical and move to Categorical Covariates box > click Plots and select Survival ten Continue > click Options and CI of exp(B) then click Continue, then OK (Figure 3.7).

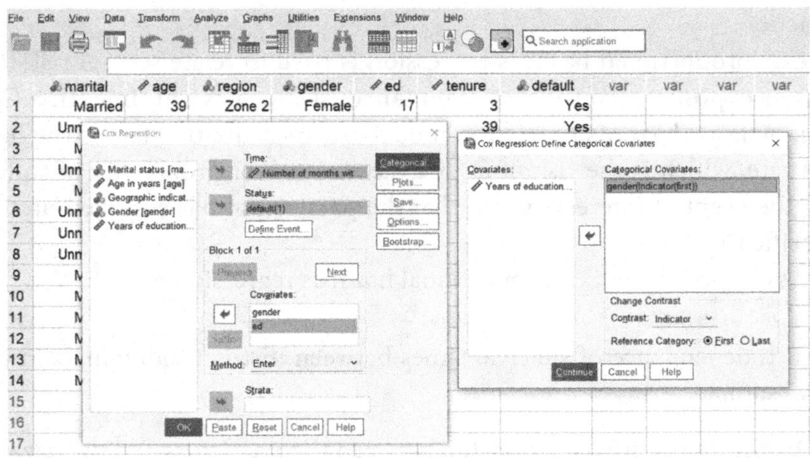

FIGURE 3.7 SPSS Procedure for Cox Regression Anlysis

SPSS Output will give the results of Cox regression model – omnibus test for coefficients (Figure 3.8) and survival curve (Figure 3.9).

Block 1: Method = Enter

Omnibus Tests of Model Coefficients[a]

-2 Log Likelihood	Overall (score)			Change From Previous Step			Change From Previous Block		
	Chi-square	df	Sig.	Chi-square	df	Sig.	Chi-square	df	Sig.
17.211	1.996	1	.158	1.986	1	.159	1.986	1	.159

a. Beginning Block Number 1. Method = Enter

Variables in the Equation

	B	SE	Wald	df	Sig.	Exp(B)	95.0% CI for Exp(B) Lower	95.0% CI for Exp(B) Upper
Gender	1.508	1.165	1.676	1	.195	4.519	.461	44.327

Covariate Means

	Mean
Gender	.500

FIGURE 3.8 SPSS Output for Cox Regression Model

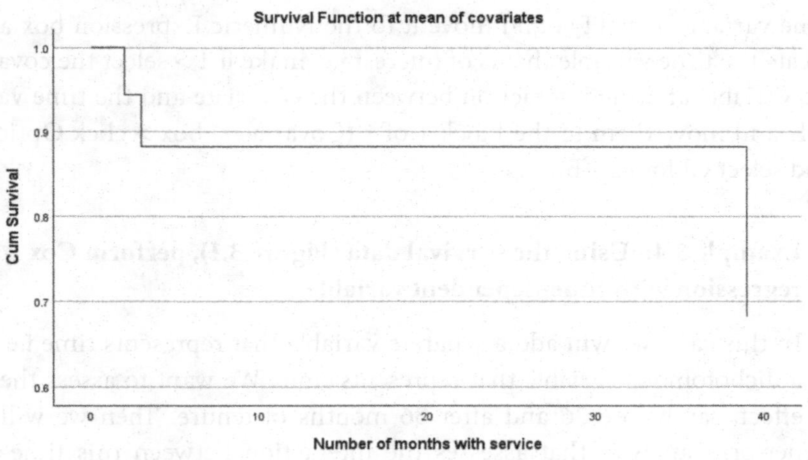

FIGURE 3.9 SPSS Output for Cox Regression Survival Curve

3.5 COX REGRESSION WITH TIME-DEPENDENT COVARIATES

The Cox regression assumes that the covariate effects on a hazard function are the same for different factor levels for all time points. This assumption may not be satisfied. The Cox regression with time-dependent covariates is a technique for modeling survival time with time-dependent covariates.

Used when the risk factors or predictors change over time.

The Cox proportional hazards regression model with time-dependent covariates takes the form:

$$ln\left\{\frac{h(t\,|\,\mathbf{x}(t))}{h_0(t)}\right\} = \beta_1 X_1(t) + \beta_2 X_2(t) + \ldots + \beta_p X_p(t).$$

where:

$h(t\,|\,\mathbf{x}(t))$ is the hazard function at time t for an individual with covariates $\mathbf{x}(\boldsymbol{t})$.

$\mathbf{x}(\boldsymbol{t})$ is a vectors of covariates at time t.

$\boldsymbol{\beta}$ is a vectors of regression coefficients corresponding to the covariates.

We usually have the partial likelihood function and use it to generate the log-partial likelihood function. Maximizing the log-partial likelihood function with respect to $\boldsymbol{\beta}$ gives the estimates of the regression coefficients.

To generate cox regression model with time-dependent covariate in SPSS, we go to: Analyze > Survival > Cox w/Time-Dep Cov > select the time variable Time[T_] and move it to the Numeric Expression box and create the time variable that is of interest i.e. make it 1 > select the covariate variable and the interaction between the covariate and the time variable and move them to the Block 1 of 1 (Covariates) box > click Options and select CI for exp(B).

Example 3.4: Using the survival data (Figure 3.1), perform Cox regression with time-dependent variable.

In this case, we will add a separate variable that represents time i.e. a dichotomous variable that represents time. We want to assess the effect, say before 36 and after 36 months of tenure. Then we will perform analysis that assesses the interaction between this time-dependent variable with the status.

We go to: Analyze > Survival > Cox w/Time-Dep Cov > select the time variable Time[T_] and move it to the Numeric Expression box and type T_>36 > click Model then move tenure to Time box and default to Status box > we want to know the effect of gender and so move gender to Covariates box. In addition, we want to create an interaction term between the time variable and gender. Select both time and gender and then click >a*b> box. It will be moved to the Block 1 of 1 (Covariates) box > click Options and select CI for exp(B) click Continue then OK (Figure 3.10).

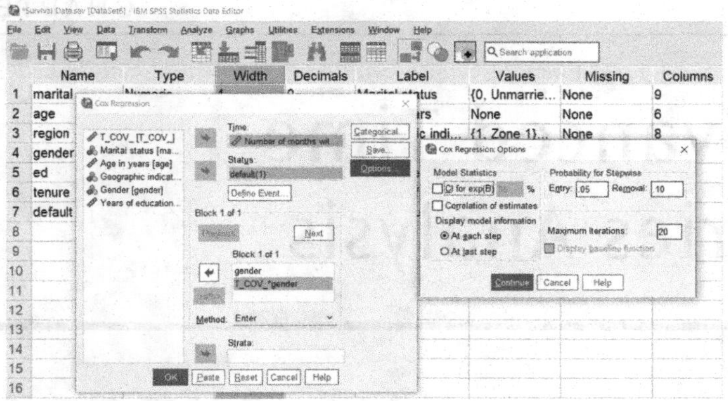

FIGURE 3.10 SPSS procedure for Cox Regression with Time-Dependent

SPSS Output will give the results for Cox regression with time-dependent variable as shown in Figure 3.11.

Omnibus Tests of Model Coefficients[a]

-2 Log Likelihood	Overall (score)			Change From Previous Step			Change From Previous Block		
	Chi-square	df	Sig.	Chi-square	df	Sig.	Chi-square	df	Sig.
16.029	2.413	2	.299	3.168	2	.205	3.168	2	.205

a. Beginning Block Number 1. Method = Enter

Variables in the Equation

	B	SE	Wald	df	Sig.	Exp(B)
Gender	11.286	191.823	.003	1	.953	79722.747
T_COV_*Gender	-10.593	191.828	.003	1	.956	.000

Covariate Means

	Mean
T_COV_	.250
Gender	.444
T_COV_*Gender	.083

FIGURE 3.11 SPSS output for Cox Regression with Time-Dependent

The findings will show the effect of gender on the tenure after 36 months.

3.6 PRACTICE EXERCISE

1. Classify the different survival data analysis techniques.
2. Using relevant SPSS data, perform and interpret the following analysis:
    a. Life table.
    b. Kaplan-Meier.
    c. Cox regression.
    d. Cox-regression with time-dependent covariate.

CHAPTER 4

Advanced Time Series Analysis

4.1 INTRODUCTION

Time series data are data that have been collected sequentially overtime. A time series is a sequence of measurements collected over time, and examples include stock price, temperature series etc. There are two main approaches used to analyze time series; *in the time domain* or *in the frequency domain.*

Time series models are used for a variety of reasons including predicting future outcomes and understanding past outcomes/trends. Time series modeling approaches can be split into: either time-domain or frequency, univariate or multivariate and linear or nonlinear.

The time-domain approach models future values as a function of past values and present values; they include autoregressive moving average models (ARMA), autoregressive integrated moving average (ARIMA) models, vector autoregressive models (VAR) and generalized autoregressive conditional heteroscedasticity (GARCH). Frequency domain models are based on the idea that time series can be represented as a function of time using sines and cosines. Frequency domain models utilize regressions on sines and cosines, rather than past and present values, to model the behavior of the data. They include spectral analysis, band spectrum regression, Fourier transform methods and spectral factorization.

Univariate time series models are models that are used when the dependent variable is a single time series. They include univariate GARCH,

DOI: 10.1201/9781003386636-4

seasonal ARIMA (SARIMA) models and univariate unit root tests. Multivariate time series models are used when there are multiple dependent variables. In addition to depending on their own past values, each series may depend on past and present values of the other series. Time series models can also be split into linear and nonlinear models depending on the presence of structural breaks in the time series data.

4.2 SEASONAL MODELS

Seasonality is a characteristic of a time series in which the data experience regular and predictable changes that recur every calendar year. Seasonal time series exhibits cyclical or periodic behavior. Seasonality in a time series is a regular pattern of changes that repeats over S time periods, where S defines the number of time periods until the pattern repeats again.

4.2.1 Differencing

Seasonality usually causes the series to be nonstationary because the average values at some particular times within the seasonal span (months, for example) may be different than the average values at other times. It may be necessary to examine differenced data when we have seasonality. *Seasonal differencing* is the difference between a value and a value with lag that is a multiple of S. Seasonal differencing removes seasonal trend and can also get rid of a seasonal random walk type of nonstationarity.

4.2.1.1 Non-seasonal Differencing

If trend is present in the data, we may also need non-seasonal differencing where the first difference (non-seasonal) will "detrend" the data.

4.2.1.2 Differencing for Trend and Seasonality

When both trend and seasonality are present, we may need to apply both a non-seasonal first difference and a seasonal difference.

4.2.2 Seasonal ARIMA Model

The seasonal ARIMA model incorporates both non-seasonal and seasonal factors in a multiplicative model. It is denoted by

$$ARIMA(p, d, q)\times(P, D, Q)S.$$

where p = non-seasonal AR order, d = non-seasonal differencing, q = non-seasonal MA order, P = seasonal AR order, D = seasonal differencing, Q = seasonal MA order, and S = time span of repeating seasonal pattern.

The general form of the SARIMA model is

$$\Phi_p(B)\Phi_P(B^s)(1-B)^d(1-B)^D Y_t = \Theta_q(B)\Theta_Q(B^s)\epsilon_t.$$

where:

$\Phi_p(B)$ is the non-seasonal autoregressive (AR) operator.
$\Phi_P(B^s)$ is the seasonal AR operator.
$\Theta_q(B)$ is the non-seasonal moving average (MA) operator.
$\Theta_Q(B^s)$ is the seasonal MA operator.
B is the backshift operator.
d is the non-seasonal differencing order.
D is the seasonal differencing order.
ϵ_t is the white noise error term.

4.2.2.1 Steps for Identifying a Seasonal Model

Step 1: Plot a time series plot of the data and examine it for features such as trend and seasonality.

Step 2: Perform any necessary differencing. If there is seasonality and no trend, then take a difference of lag S.

Step 3: Examine the Autocorrelation Function (ACF) and Partial Autocorrelation Function (PACF) of the differenced data (if differencing is necessary).

- **Non-seasonal terms**: Examine the early lags (1, 2, 3, …) to judge non-seasonal terms. Spikes in the ACF (at low lags) with a tapering PACF indicate non-seasonal MA terms. Spikes in the PACF (at low lags) with a tapering ACF indicate possible non-seasonal AR terms.

- **Seasonal terms**: Examine the patterns across lags that are multiples of S. For example, for monthly data, look at lags 12, 24, 36, and so on (probably won't need to look at much more than the first two or three seasonal multiples). Judge the ACF and PACF at the seasonal lags in the same way you do for the earlier lags.

Step 4: Estimate the model(s) that might be reasonable on the basis of Step 3. Don't forget to include any differencing that you did before looking at the ACF and PACF.

Step 5: Examine the residuals (with ACF, Box-Pierce and any other means) to see if the model seems good. Compare Akaike Information Criterion (AIC) or Bayesian Information Criterion (BIC) values to choose among several models. If things don't look good here, it's back to Step 3 (or maybe even Step 2).

To identify seasonal models in SPSS, we go to: Analyze > Forecasting > Create Traditional Models > Time Series Modeler.

We first define time series data by: Data > Define date and time then select Years, months

We can then forecast with seasonal model.

Example 4.1: Using the time series data (Figure 4.1.), forecast using SARIMA.

We first define time series data by: Data > Define date and time then select Years, months. Edit Years to start from 2010 and months to start from 1 (January) as shown in Figure 4.2.

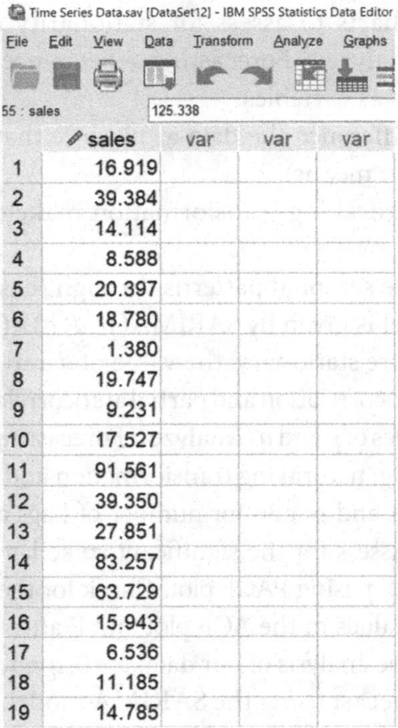

Time Series Data.sav [DataSet12] - IBM SPSS Statistics Data Editor

File Edit View Data Transform Analyze Graphs

55 : sales 125.338

	sales	var	var	var
1	16.919			
2	39.384			
3	14.114			
4	8.588			
5	20.397			
6	18.780			
7	1.380			
8	19.747			
9	9.231			
10	17.527			
11	91.561			
12	39.350			
13	27.851			
14	83.257			
15	63.729			
16	15.943			
17	6.536			
18	11.185			
19	14.785			

FIGURE 4.1 Time Series Data

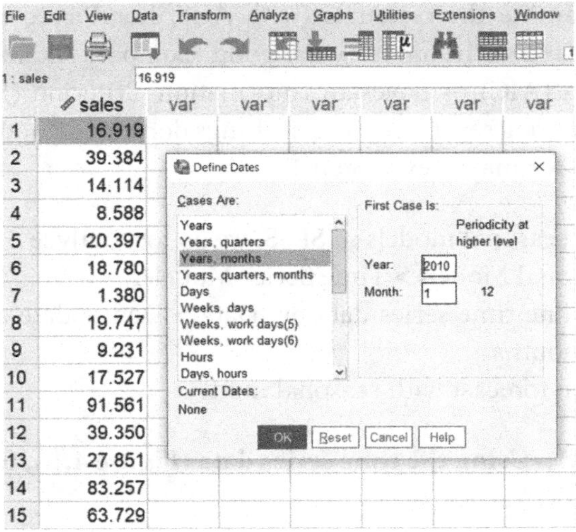

FIGURE 4.2 SPSS Procedure for Defining Data and Time

We can first have the visual representation of the time series data using sequence charts to assess for stationarity and trends in the data. We go to Analyze > Forecasting > Sequence Charts and select sales in thousands as variables.

We can then difference the data as observe that the data are now stationary (i.e. zero means).

We then do natural log transformation make the variance to be constant.

We then remove seasonal patterns through seasonal differencing.

SARIMA model is given by SARIMA$(p, d, q)\times(P, D, Q)_{12}$

Since the data are stationary, the value of d and D =1.

We generate autocorrelation and partial autocorrelation plots to be able to estimate the values of p and q (Analyze > Forecasting > Autocorrelation > check differencing, natural log transformation and seasonal differencing > click Options and put 36 for number of Lags and select Bartlett's Approximation). Assess for the significant spike lag for the values of q using ACF plot and q using PACF plot. Check for the exponential decay after lag 12 for Q values in the ACF plot and P after lag 12 in the PACF plot. Hence from the analysis of our data, $p=0$, $q=1$, $P=1$ and $Q=0$.

We can now forecast using the SARIMA model. We go to Analyze > Forecasting > Create Traditional Models > select sales and drag it to Dependent Variable box > under Method select ARIMA and click Criteria then fill in the values for $p=0$, $d=0$, $q=0$ for Non-seasonal and $P = 1$, $D = 1$ and $Q = 0$ for Seasonal and then select Natural Log transformation then click Continue as shown in Figure 4.3.

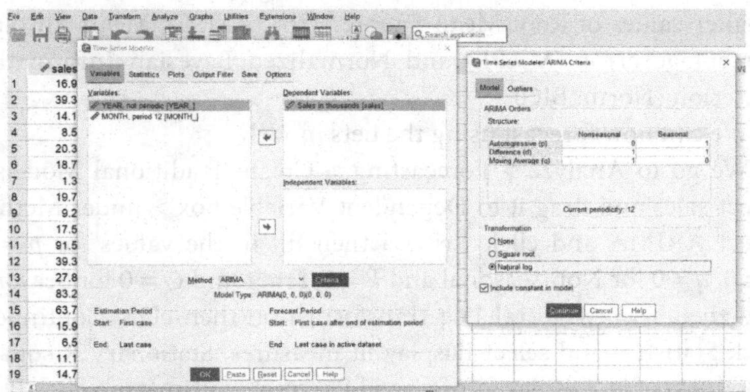

FIGURE 4.3 SPSS Procedure for Forecasting Using SARIMA Model

Click Statistics and select R Square, Root mean square error, Mean absolute percentage error, Normalized BIC, Goodness of fit and Parameter estimates.

Click Plots and select Series, Observed values, forecast, Residual autocorrelation function (ACF) and Residual partial autocorrelation function (PACF) then click OK. SPSS will give output for the estimated SARIMA model as shown in Figure 4.4.

Model Description

			Model Type
Model ID	Sales in thousands	Model_1	ARIMA(0,1,1)(1,1,0)

Model Summary

Model Fit

Fit Statistic	Mean	SE	Minimum	Maximum	Percentile 5	10	25	50	75	90	95
Stationary R-squared	.457		.457	.457	.457	.457	.457	.457	.457	.457	.457
R-squared	-17.128		-17.128	-17.128	-17.128	-17.128	-17.128	-17.128	-17.128	-17.128	-17.128
RMSE	339.423		339.423	339.423	339.423	339.423	339.423	339.423	339.423	339.423	339.423
MAPE	1083.872		1083.872	1083.872	1083.872	1083.872	1083.872	1083.872	1083.872	1083.872	1083.872
MaxAPE	22923.915		22923.915	22923.915	22923.915	22923.915	22923.915	22923.915	22923.915	22923.915	22923.915
MAE	176.048		176.048	176.048	176.048	176.048	176.048	176.048	176.048	176.048	176.048
MaxAE	1630.704		1630.704	1630.704	1630.704	1630.704	1630.704	1630.704	1630.704	1630.704	1630.704
Normalized BIC	11.797		11.797	11.797	11.797	11.797	11.797	11.797	11.797	11.797	11.797

Model Statistics

Model	Number of Predictors	Model Fit statistics: Stationary R-squared	R-squared	RMSE	MAPE	Normalized BIC	Ljung-Box Q(18): Statistics	DF	Sig.	Number of Outliers
Sales in thousands-Model_1	0	.457	-17.128	339.423	1083.872	11.797	6.772	16	.977	<.001

ARIMA Model Parameters

					Estimate	SE	t	Sig.
Sales in thousands-Model_1	Sales in thousands	Natural Logarithm	Constant		-.001	.025	-.035	.972
			Difference		1			
			MA	Lag 1	.797	.066	12.011	<.001
			AR, Seasonal	Lag 1	-.354	.109	-3.239	.002
			Seasonal Difference		1			

FIGURE 4.4 SPSS Output for the Estimated SARIMA Model

You may create another SARIMA models by changing the values of p, d, q and P, D, Q for comparison so that you can get the best model by looking at the model with the highest value of R-squared, smaller values of Root Mean Square error (RMSE), Mean Absolute Percentage error (MAPE) and Normalized Bayesian Information Criterion (Norm. BIC).

We can now forecast using the bets model.

We go to Analyze > Forecasting > Create Traditional Models > select sales and drag it to Dependent Variable box > under Method select ARIMA and click Criteria then fill in the values for $p = 0$, $d = 0$, $q = 0$ for Non-seasonal and $P = 1$, $D = 1$ and $Q = 0$ for Seasonal and then select Natural Log transformation then click Continue > click Statistics and select Display fit measures, Stationary R square, Goodness of fit, Display forecast and Parameter estimates > click Plots and select Series, Observed values and Forecast > click Options and select First case after the end of estimation period through a specified date. Put the period you would like to forecast e.g. 6 months after Jan 2029 (write 2019 for Year and 6 for Months) as shown in Figure 4.5.

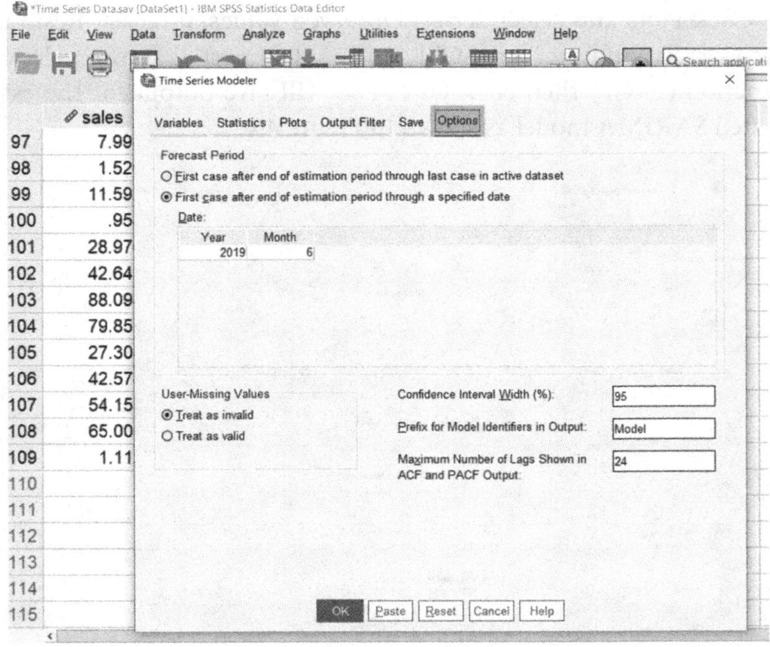

FIGURE 4.5 SPSS Procedure for Forecasting for the Next 6 Months

SPSS Output will show the forecast values for the next 6 months (Figure 4.6).

Forecast

Model		Feb 2019	Mar 2019	Apr 2019	May 2019	Jun 2019
Sales in thousands-Model_1	Forecast	24.208	65.789	17.055	226.200	323.879
	UCL	155.929	429.734	112.837	1514.099	2190.973
	LCL	.374	.912	.212	2.528	3.257

For each model, forecasts start after the last non-missing in the range of the requested estimation period, and end at the last period for which non-missing values of all the predictors are available or at the end date of the requested forecast period, whichever is earlier.

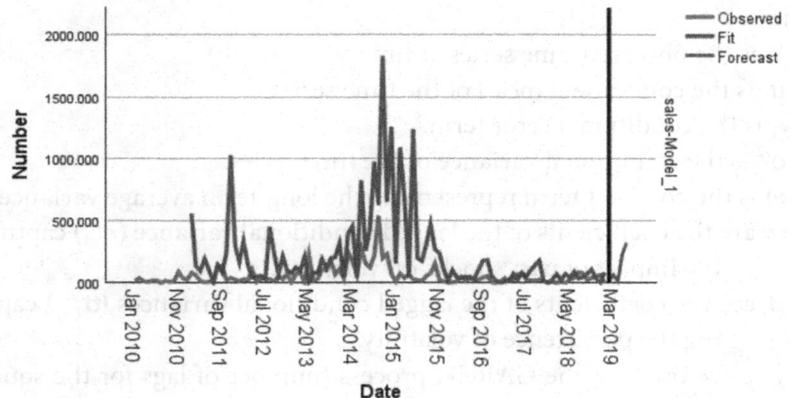

FIGURE 4.6 SPSS Output Displaying Forecast for the Next 6 Months

4.3 ADVANCED TIME SERIES MODELS

We will examine models for periods of volatile variance (Autoregressive Conditional Heteroscedasticity, i.e., ARCH models).

4.3.1 ARCH/GARCH Models

An ARCH model is a model for the variance of a time series. ARCH models are used to describe a changing, possibly volatile variance. ARCH models were created in the context of econometric and finance problems having to do with the amount that investments or stocks increase (or decrease) per time period, so there's a tendency to describe them as models for that type of variable. An ARCH model could be used for any series that has periods of increased or decreased variance. The GARCH model is a time series model used to describe and predict the volatility of financial returns or other time series data.

In general, An ARCH(m) process is one for which the variance at time *t* is conditional on observations at the previous m times.

The basic GARCH(p,q) model is given by

$$Y_t = \mu_t + \epsilon_t.$$

$$\epsilon_t = \sigma_t \varepsilon_t.$$

$$\sigma_t^2 = \omega + \sum_{i=1}^{p} \alpha_i \epsilon_{t-1}^2 + \sum_{j=1}^{q} \beta_j \sigma_{t-j}^2.$$

where:

Y_t is the observed time series at time *t*.

μ_t is the conditional mean of the time series.

ϵ_t is the conditional error term.

σ_t^2 is the conditional variance of the time series.

ω is the constant term representing the long-term average variance.

α_i are the coefficients of the lagged conditional variance (ϵ_{t-1}^2) capturing the impact of past shocks on volatility.

β_j are the coefficients of the lagged conditional variances (σ_{t-j}^2) capturing the persistence of volatility.

p is the order of the GARCH process (number of lags for the squared residuals).

q is the order of the ARCH process (number of lags for the conditional variance).

The parameters of the GARCH model (ω, α_i, β_j) are estimated using maximum likelihood estimation (MLE).

The ARCH model assumes that the conditional mean of the error term in a time series model is constant (zero), but its conditional variance is not.

A GARCH model uses values of the past squared observations and past variances to model the variance at time *t*.

4.3.1.1 Identifying an ARCH/GARCH Model

The best identification tool may be a time series plot of the series. It's usually easy to spot periods of increased variation sprinkled through the series. It can be fruitful to look at the ACF and PACF of both y_t and y_t^2.

To generate an ARCH/GARCH model in SPSS, we use STATS GARCH extension which is installed as part of IBM SPSS Statistics-Essentials for R.

To download and install GARCH extension, we go to Extensions > Extension Hub > Search for the name of the extension i.e. STATS GARCH > Tick Get Extension then click OK > then accept the terms and conditions, then click Finish to install GARCH extension.

To generate GARCH Model in SPSS we go to Analyze > Forecasting > GARCH > select the variables you want to include in the model, i.e., variable to Model, Variance Model Regressors, Mean Model Regressors and ID variable > click Model to specify the GARCH order (p and q parameters), then click Continue then OK.

The basic GARCH model can be extended to Exponential GARCH (EGARCH) to allow for symmetric effects of positive and negative shocks on volatility. It can also be extended to Integrated GARCH (IGARCH) to allow for time-varying volatility persistence.

Other advanced time series models include State Space Models like the Kalman Filter, Dynamic Linear Models, and Bayesian Structural Time Series, among others.

4.4 MULTIVARIATE TIME SERIES ANALYSIS

Multivariate time series analysis involves the analysis of data over time that consists of multiple interdependent variables. A multivariate time series has more than one time series variable. Each variable depends not only on its past values but also has some dependency on other variables. One of the most commonly used methods for multivariate time series forecasting is the vector auto regression (VAR). In VAR, each variable is a linear function of the past values of itself and the past values of all the other variables. In this case, we will examine AR models for multivariate time series.

4.4.1 Vector Autoregressive [VAR(p)] Models

Vector autoregressive (VAR) models are used for multivariate time series. The structure is that each variable is a linear function of past lags of itself and past lags of the other variables. In general, for a VAR(p) model, the first p lags of each variable in the system would be used as regression predictors for each variable. VAR models are a specific case of more general VARMA models. VARMA models for multivariate time series include the VAR structure above along with moving average terms for each variable. The VAR model is a multivariate time series model that captures the dynamic relationships among multiple variables. It extends the concept of univariate autoregressive models to multiple time series variables simultaneously.

The VAR(p) model of order p for k variables is given by

$$\mathbf{Y}_t = \mathbf{c} + \mathbf{A}_1 \mathbf{Y}_{t-1} + \mathbf{A}_2 \mathbf{Y}_{t-2} + \ldots + \mathbf{A}_p \mathbf{Y}_{t-p} + \boldsymbol{\varepsilon}_t.$$

where:

$\mathbf{Y}_t$ is a $k \times 1$ vectors of variables at time t.
$\mathbf{c}$ is $k \times 1$ vectors of intercept terms.
$\mathbf{A}_i$ for $i = 1, 2, \ldots p$ are $k \times 1$ coefficient matrices representing the lagged effects of the variables up to order p.
ε_t is $k \times 1$ vector of error terms at time t.

Estimation of the parameters c and $\mathbf{A}_i$ in the VAR model is done using ordinary least squares (OLS), maximum likelihood estimation (MLE), or Bayesian methods.

4.4.2 Granger's Causality Test

Granger's causality test can be used to identify the relationship between variables prior to model building. This is important because if there is no relationship between variables, they can be excluded and modeled separately. Conversely, if a relationship exists, the variables must be considered in the modeling phase. The VAR model allows for testing causal relationships among variables using the Granger causality test.

The null hypothesis for the Granger causality test is that the lagged values of the potential causal variable(s) do not provide significant predictive power for the variable of interest. The test statistic for the Granger causality test follows an F-distribution and is computed based on the sum of squared errors from the two regression models, i.e., the restricted model (without casual variables) and the unrestricted model (with casual variables), and is given by:

The restricted model is given by

$$Y_t = \alpha + \sum_{i=1}^{p} \beta_i Y_{t-i} + \epsilon_t.$$

The unrestricted model is given by

$$Y_t = \alpha + \sum_{i=1}^{p} \beta_i Y_{t-i} + \sum_{j=1}^{q} \gamma_j X_{t-j} + \epsilon_t.$$

where:

Y_t is the variable of interest.

X_t is the potential causal variable.

p and q are the lag orders for Y_t and X_t, respectively.

α, β_i and γ_j are coefficients to be estimated.

ϵ_t is the error term.

The F-test is given by

$$F = \frac{(ESS_R - ESS_U)/q}{(ESS_U)/(n-p-q-1)}.$$

where:

ESS_R is the explained sum of squares from the restricted model.

ESS_U is the explained sum of squares from the unrestricted model.

q is the number of parameters in the unrestricted model.

n is the number of observations.

p is the number of lagged terms included in the models.

VAR models can be used for forecasting and making predictions about the future values of the variables in the system.

To generate a VAR model in SPSS: SPSS does not have procedures for performing Granger's Causality Test. We can manually write the codes/model in SPSS Syntax.

Other statistical software like R, Python, Eviews and STATA can be used to perform the test.

To generate a VAR model in SPSS, we go to: SPSS does not have procedures for performing Covariance-Stationary Vector Processes which VAR is part of. We use SPSS Syntax window to write the codes for VAR model.

Other statistical software such as STATA, Eviews, Python, and R can be used to run the VAR Model.

4.4.2.1 Stationarity of a Multivariate Time Series

We can use Johansen's test for checking the stationarity of any multivariate time series data. There are other techniques that can be used including Augmented Dickey-Fuller (ADF) Test, Differencing and Multivariate Unit Root Tests among others.

The Johansen test statistic follows a Chi-square distribution and is based on the trace and maximum eigenvalue tests, and it involves estimating a vector error correction model (VECM) and performing likelihood ratio tests to determine the number of cointegrating vectors.

4.5 NONSTATIONARY TIME SERIES ANALYSIS

Nonstationary time series is whereby values and associations between and among variables do vary with time. Nonstationary time series are time series whose (statistical) properties change over time. Non-stationarity can arise due to trends, seasonality, or other systematic patterns present in the data. Non-stationary time series exhibits trends. It has seasonal patterns and non-constant variance (heteroscedasticity) over time. The autocorrelation structure of non-stationary time series may change over time.

Non-stationarity can be deterministic trends, stochastic trends, or seasonal non-stationarity. There are several techniques for handling non-stationarity, and they include detrending, seasonal adjustment, differencing, transformation, and using model that account for non-stationarity like seasonal ARIMA or structural time series models.

A process is said to be stationary when its joint probability distribution does not change when shifted in time. However, stationarity is a strong assumption and can be violated in practice.

4.5.1 Unit Roots and Random Walks

A random walk is a time series in which the value of the series in one period is equivalent to the value of the series in the previous period plus the unforeseeable random error.

Consider an AR(1) model. If the time series originates from an AR(1) model, then the time series is covariance stationary if the absolute value of the lag coefficient is less than 1. When the lag coefficient is precisely equal to 1, then the time series is said to have a unit root. In other words, the time series is a random walk and hence not covariance stationary.

4.5.1.1 Unit Root Test

The unit root test is done using the augmented Dickey-Fuller (ADF) test. The test involves OLS estimation of the parameters where the difference of the time series is regressed on the lagged level, appropriate deterministic terms and the lagged difference. The unit root problem can also be expressed using the lag polynomial. The test is conducted under the null hypothesis that the series has a unit root, indicating non-stationarity and it follows a *t*-distribution.

The ADF test statistic is computed by regressing the differenced series (Δy_t) on lagged values of the original series (y_{t-1}) and possibly other lagged differences of the series.

4.5.2 Cointegration and Error Correction Models

Cointegration essentially means two time series have a long-run relationship. Two series are cointegrated when their trends are not too far apart and are in some sense similar. A set of k time series variables $Y_1, Y_2, \ldots, Y_k$ are said to be cointegrated if there exists a linear combination of these variables that is stationary.

The cointegration equation for k cointegrated variables is given by

$$Y_t = \alpha_0 + \alpha_1 Y_{t-1} + \alpha_2 Y_{t-2} + \ldots + \alpha_{k-1} Y_{t-(k-1)} + \epsilon_t .$$

where

Y_t is the vector of cointegrated variable at time t.

α_0 is the constant term.

α_i are the coefficients of the lagged cointegrated variables.

ϵ_t is the error term.

The error-correction model (ECM) is a form of multiple time series models most commonly used for data where the underlying variables have a long-run common stochastic trend or cointegration. The ECM is a modeling framework used to describe the short-term dynamics of cointegrated variables.

If there isn't a long-run relationship, an ECM is not appropriate.

The ECM is generally specified as a regression model that includes the first-differenced dependent variable and lagged error correction terms and is given by

$$\Delta Y_t = \beta_0 + \beta_1 \Delta Y_{t-1} + \gamma\left(Y_{t-1} - \hat{Y}_{t-1}\right) + \sum_{i=1}^{p} \beta_i \Delta Y_{t-i} + \epsilon_t .$$

where

ΔY_t is the first-differenced dependent variable.

β_0 is the constant term.

β_i are the coefficients of the lagged differenced variables.

γ is the coefficient of the error correction term $(Y_{t-1} - \hat{Y}_{t-1})$, where $\hat{Y}_{t-1}$ is the predicted values of Y_{t-1} from the cointegration equation.
p is the lag order for the model.
ϵ_t is the error term.

To generate the error-correction model in SPSS: SPSS doesn't have procedures for estimating vector error correction models (VECM). Other statistical software like R, Eviews and STATA can be used to generate VECM.

We use the linear regression path in SPSS i.e. Analyze > regression > Linear then specify the first-differenced dependent variable as the dependent variable and include lagged first-differenced variables, lagged levels of the variables, and the error correction term as predictors.

4.6 TIME SERIES ANALYSIS IN FREQUENCY DOMAIN

Time series analysis in the frequency domain involves examining the frequency components of a time series data, which provides insights into the periodic patterns and cyclical behavior present in the data. There are several expression for time series analysis in the frequency domain, and they include Fourier transform, power spectrum, periodogram, and spectral density.

The Fourier transform is a mathematical tool used to decompose a time series into its frequency components where time series is represented as a sum of sine and cosine waves with different frequencies. It is expressed as

$$X(f) = \int_{-\infty}^{\infty} x(t) e^{-2\pi i f t} \, dt.$$

where:

$X(f)$ is the frequency-domanin representation of the time series.
$x(t)$ is the time-domain representation of the time series.
f is the frequency variable.
i is the imaginary unit.

The power spectrum represents the distribution of power (or variance) across different frequencies in the time series. It provides information about the relative importance of different frequency components. It is expressed as

$$P(f) = |X(f)|^2.$$

where:

$P(f)$ is the power spectrum.

$X(f)$ is the Fourier transform of the time series.

4.6.1 The Periodogram

Any time series can be expressed as a combination of cosine and sine waves with differing periods (how long it takes to complete a full cycle) and amplitudes (maximum/minimum value during the cycle). This fact can be utilized to examine the periodic (cyclical) behavior in a time series.

A periodogram is used to identify the dominant periods (or frequencies) of a time series. This can be a helpful tool for identifying the dominant cyclical behavior in a series, particularly when the cycles are not related to the commonly encountered monthly or quarterly seasonality.

Period (T) is the number of time periods required to complete a single cycle of the cosine function.

Frequency $\omega = \frac{1}{T}$ is the fraction of the complete cycle that's completed in a single time period.

The periodogram is an estimator of the power spectrum and is obtained by computing the squared magnitude of the discrete Fourier transform (DFT) of the time series. It is a useful tool for visualizing the frequency content of a time series. It is expressed as

$$I(f) = \frac{1}{N}\left|\sum_{t=1}^{N} x(t)e^{-2\pi ift}\right|^2.$$

where:

$I(f)$ is the periodogram.

N is the length of time series.

In SPSS, we generate Periodogram through: Analyze > Forecasting > Spectral Analysis > Under Plot select Periodogram.

Example 4.2: Using the time series data (Figure 4.1) perform spectral analysis.

We go to: Analyze > Forecasting > Spectral Analysis > Under Plot, select Periodogram By Frequency (Figure 4.7).

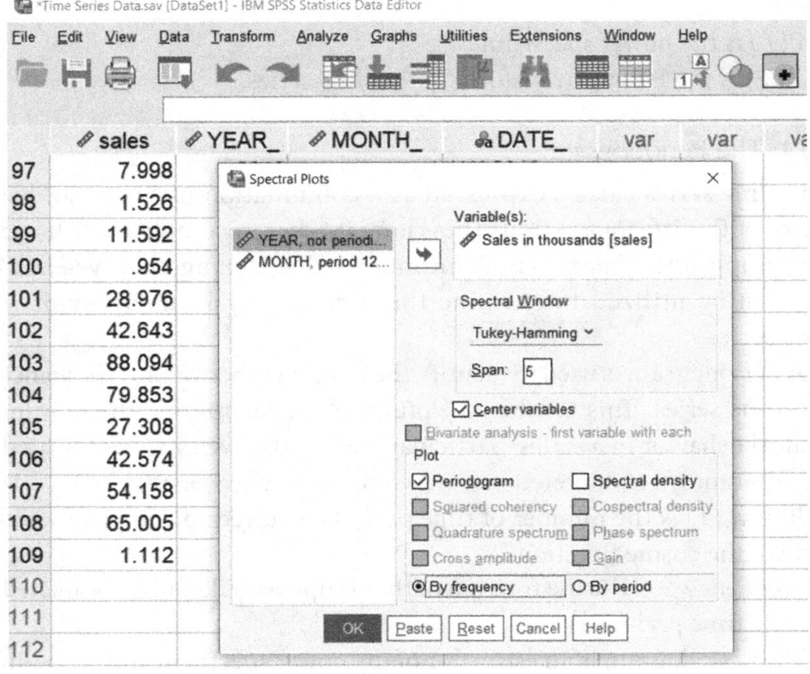

FIGURE 4.7 SPSS Procedure for Generating Periodogram

SPSS Output will give the Periodogram (Figure 4.8).

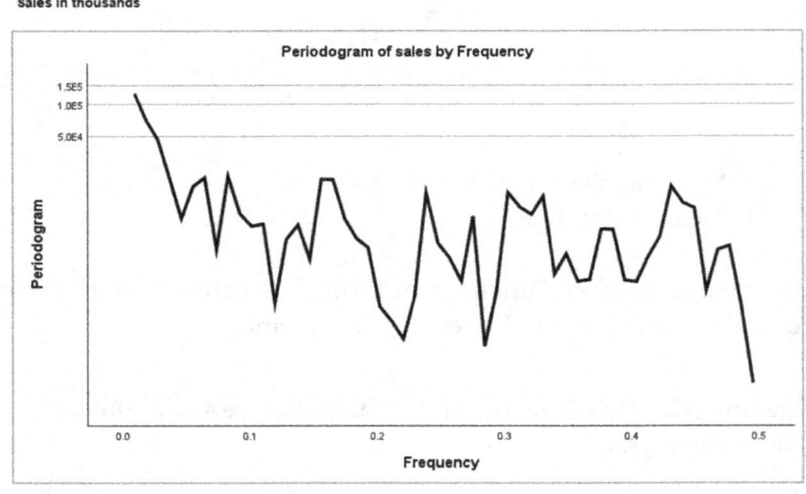

FIGURE 4.8 SPSS Output Displaying Periodogram

4.6.2 Spectral Density

Periodogram, a function/graph that displays information about the periodic components of a time series. Any time series can be expressed as a sum of cosine and sine waves oscillating at the fundamental (harmonic) frequencies. The periodogram is a sample estimate of a population function called the spectral density, which is a frequency domain characterization of a population stationary time series. The spectral density is a frequency domain representation of a time series that is directly related to the autocovariance time domain representation.

In *spectral analysis*, we view a time series as a sum of cosine waves with varying amplitudes and frequencies. One goal of an analysis is to identify the important frequencies (or periods) in the observed series. A starting tool for doing this is the periodogram. The periodogram graphs a measure of the relative importance of possible frequency values that might explain the oscillation pattern of the observed data. *Spectral analysis* is the analysis of the dominant frequencies in a time series. In practice, spectral analysis imposes smoothing techniques on the periodogram. With certain assumptions, we can also create confidence intervals to estimate the peak frequency regions. Spectral analysis can also be used to examine the association between two different time series.

The spectral density represents the distribution of power per unit frequency and is often used in the analysis of stationary time series. It is the Fourier transform of the autocovariance function of the time series. It is expressed as

$$S(f) = \int_{-\infty}^{\infty} R(T)e^{-2\pi i f T}\, dT.$$

where:

$S(f)$ is the spectral density.

$R(T)$ is the autocovariance furnction of the time series.

T is the lag variable.

4.6.2.1 Methods for Estimating the Spectral Density

The raw periodogram is a rough sample estimate of the population spectral density. The estimate is "rough", in part, because we only use the discrete fundamental harmonic frequencies for the periodogram, whereas the spectral density is defined over a continuum of frequencies.

An alternative approach to smoothing the periodogram is a parametric estimation approach based on the fact that any stationary time series can be approximated by an AR model of some order (although it might be a high order). In this approach a suitable AR model is found, and then the spectral density is estimated as the spectral density for that estimated AR model.

In SPSS, we generate Spectral Density through: Analyze > Forecasting > Spectral Analysis > Under Plot, select Spectral Density.

Example 4.3: Using the time series data (Figure 4.1), perform spectral analysis.

We go to: Analyze > Forecasting > Spectral Analysis > Under Plot select Spectral Density (Figure 4.9).

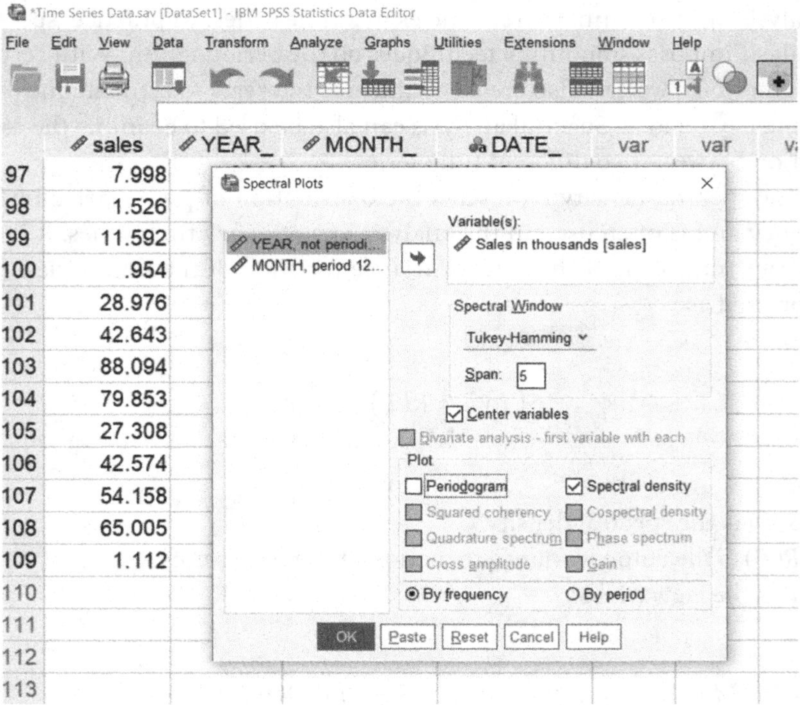

FIGURE 4.9 SPSS Procedure for Spectral Density Analysis

SPSS Output will give the results of spectral density (Figure 4.10).

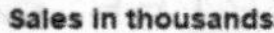

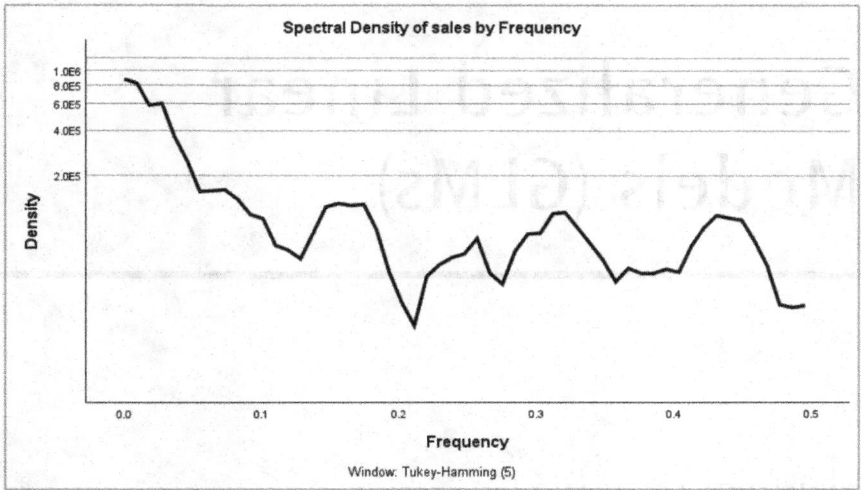

FIGURE 4.10 SPSS Output for Spectral Analysis

4.7 PRACTICE EXERCISE

1. Classify the different advanced time series analysis techniques.
2. Using relevant SPSS data, perform and interpret the following analysis:
   a. Forecast using seasonal ARIMA (SARIMA) model.
   b. ARCH/GARCH model.
   c. Vector auto regression (VAR).
   d. Error-correction model.
   e. Spectral density.

CHAPTER 5

Generalized Linear Models (GLMs)

5.1 INTRODUCTION

Generalized linear models (GLMs) refer to the models involving link functions. This is an extension of general linear model so that a dependent variable can be linearly related to factors and/or covariates by using a link function. The dependent variable does not require normal assumption. The dependent variable could be count (as in Poisson regression model or negative binomial regression model) or ordinal (as in logistic regression model). GLMs have three components: a random component, a systematic component and a link function.

When the dependent variable does not follow a nice bell-shaped Normal distribution, then we will need to use GLM. The GLM is a more general class of linear models that change the distribution of your dependent variable. The general form of a GLM can be expressed as

$$g(\mu_i) = \eta_i = \beta_0 + \beta_1 x_{i1} + \beta_2 x_{i2} + \ldots + \beta_p x_{ip}.$$

where:

μ_i is the expected values of the response variable Y_i.

η_i is the linear predictor.

$g(.)$ is the link function.

$\beta_0, \beta_1, \ldots, \beta_p$ are the coefficients to be estimated.

$x_{i1}, x_{i2}, \ldots, x_{ip}$ are the values of the explanatory variables in the i-th observation.

DOI: 10.1201/9781003386636-5

5.1.1 Exponential Family of Distributions

GLMs assume that the response variable Y belongs to the exponential family of distributions, which has the general form:

$$f_Y\left(y:\theta,\phi\right)=exp\left(\frac{y\theta-b(\theta)}{\phi}+c\left(y,\phi\right)\right).$$

where:

θ is the canonical parameter, ϕ is the dispersion parameter.

$b(\theta)$ and $c(y,\phi)$ are specific functions that define the particular distribution.

The parameters β for the GLM are estimated using maximum likelihood estimation (MLE) from where the likelihood function is constructed based on the chosen distribution and the link function, and the estimates are found by maximizing this function.

The following table shows when to use the different GLMs:

Family Name	Description
Binomial	Binary logistic regression, useful when the response is either 0 or 1.
Gaussian	Standard linear regression. Using this family will give you the same result as linear model.
Gamma	Gamma regression, useful for highly positively skewed data.
Inverse.Gaussian	Inverse-Gaussian regression, useful when the dependent variable is strictly positive and skewed to the right.
Poisson	Poisson regression, useful for count data.

In SPSS, we perform GLM analyses using the path: Analyze > Generalized Linear Models > Generalized Linear Models > we will get SPSS pop up window with the Generalized Linear Models dialogue box.

The main dialog box for GLMs technique has the following tabs:

- Type of model: Users select the type of response to be analyzed. This technique allows for scale response, ordinal response, count data, binary response and mixture data. Here, you select the model type e.g. "Poisson loglinear" for Poisson regression model. One can do multiple linear regression by selecting Linear under the scale Response. You can select binary logistic and a custom model. Among the custom models are normal, binomial and negative binomial. However, note that SPSS does not estimate the scale parameter under the negative binomial model.

- Response: This menu asks for dependent variable and scale weight (if any). Select your dependent variable. If the response variable is binary, you can specify the reference category. You can also specify a binary response in the form of "*n*" and "*y*", where "*n*" is the number of trials and "*y*" is the number of "events" or "successes".
- Predictors: This menu asks for factors and covariates to be used as independent variables. Select the factors, covariates and any offset variable you may have. "Options" in this submenu will allow you to specify how to handle missing values and how to order the factors.
- Model: This menu asks users to define the model. Specify the model. The default model is not the full factorial. It is the intercept-only model. You must specify the model terms explicitly.
- Estimation: This allows users to decide how the model parameters are to be estimated, and how the covariance matrix required for the modeling to be estimated. You can also make changes to the default settings for the iterations.
- Statistics: Users can choose a variety of statistics to be reported in the output. By default, Type III sum of squares are computed. You can change this.
- Expectation-Maximization (EM) means: This tab will allow users to display the estimated marginal means for the factor levels.
- Save: Users can save various statistics for further analysis and model diagnostics.
- Export: Users can export the final model as data or as XML.

In SPSS, we go to: Analyze > Generalized Linear Models > Generalized Linear Models.

5.2 MODELS FOR PROPORTIONS

5.2.1 Logistic Regression

Logistic Regression is used when the dependent variable (target) is categorical. The main task in logistic regression analysis is to estimate the log odds of an event.

There are three types of logistic regression:

1. Binary logistic regression: The categorical response has only two possible outcomes. Example: spam or not.
2. Multinomial logistic regression: Three or more categories without ordering. Example: Predicting which food is preferred more (veg, non-veg and vegan).
3. Ordinal logistic regression: Three or more categories with ordering. Example: Movie rating from 1 to 5.

5.2.1.1 Binary Logistic Regression

Binary (also called binomial) logistic regression is appropriate when the outcome is a dichotomous variable (i.e. categorical with only two categories), and the predictors are of any type: nominal, ordinal and/or interval/ratio (numeric).

The log transformation of the *p* values to a log distribution enables us to create a link with the normal regression equation. The log distribution (or logistic transformation of *p*) is also called the logit of *p* or logit(*p*).

In logistic regression, a logistic transformation of the odds (referred to as logit) serves as the depending variable:

$$\log(odds) = logit(P) = ln\left(\frac{P}{1-P}\right).$$

If we take the above dependent variable and add a regression equation for the independent variables, we get a logistic regression:

$$logit(P) = \log\left(\frac{\mu_i}{1-\mu_i}\right) = \beta_0 + \beta_1 x_1 + \beta_2 x_2 + \ldots + \beta_k x_k.$$

We can calculate *P* as

$$P = \frac{\exp(\beta_0 + \beta_1 x_1 + \beta_2 x_2 + \ldots + \beta_k x_k)}{1+\exp(\beta_0 + \beta_1 x_1 + \beta_2 x_2 + \ldots + \beta_k x_k)} = \frac{\exp(X\beta)}{1+\exp(X\beta)}.$$

where:

P = the probability that a case is in a particular category.
μ_i is the probability that response variable Y_i equals to 1.
exp = the exponential function (approx. 2.72).

β_0 = the constant (or intercept) of the equation.
$\beta_1, \beta_2, \ldots \beta_k$ = the coefficient (or slope) of the predictor variables.

5.2.1.2 Logits or Log Odds

Odds value can range from 0 to infinity and tell you how much more likely it is that an observation is a member of the target group rather than a member of the other group.

$$Odds = \frac{P}{1-P}.$$

If the probability is 0.80, the odds are 4 to 1 or 0.80/0.20; if the probability is 0.25, the odds are 0.33 (0.25/0.75).

The odds ratio (OR) estimates the change in the odds of membership in the target group for a one unit increase in the predictor. It is calculated by using the regression coefficient of the predictor as the exponent or exp.

For example, lets predict accountancy success by a maths competency predictor b = 2.69. Thus, the odds ratio is exp(2.69) or 14.73. Therefore, the odds of passing are 14.73 times greater for a student for example who had a pretest score of 5 than for a student whose pretest score was 4.

In logistic regression, hypotheses are of interest:

H_0: All the coefficients in the regression equation take the value zero.

H_1: The current model under consideration is accurate and differs significantly from the null of zero, that is, significantly better than the chance or random prediction level of the null hypothesis.

To evaluate the hypothesis, we work out the likelihood of observing the data we actually did observe under each of these hypotheses.

To perform Binary Logistic Regression in SPSS, we go to: Analyze > Regression > Binary Logistic > select the outcome variable (categorical) and move it to the Dependent: box. Then select the predictor variables (numeric variables) and move them to the Covariates: box. If one or more of the predictors were categorical, we would need to click on the Categorical button to specify them > Click Options and select Classification plots, Hosmer-Lemeshow goodness-of-fit, Casewise listing of residuals, Correlations of estimates, Iteration history and CI for exp(B) then click the Continue then OK. SPSS will display the results for Binary Logistic Regression.

Example 5.1: Using binary logistic data (Figure 5.1), perform binary logistic regression analysis using one continuous and one dichotomous predictor variable plus one dichotomous dependent variable.

Binary Logistic Data.sav [DataSet3] - IBM SPSS Statistics Data Editor

File Edit View Data Transform Analyze Graphs Utilities Extens

83 : gender

	age	gender	union	var	var
1	39	Female	No		
2	43	Male	No		
3	49	Male	No		
4	35	Male	No		
5	32	Male	No		
6	21	Female	No		
7	79	Female	No		
8	64	Male	No		
9	62	Male	No		
10	61	Male	Yes		
11	48	Female	No		
12	35	Female	No		
13	25	Female	No		
14	61	Female	No		
15	57	Male	No		
16	52	Male	No		
17	19	Female	Yes		
18	65	Male	No		
19	22	Female	No		

FIGURE 5.1 Binary Logistics Data

We go to: Analyze > Regression > Binary Logistic > select the dependent variable (union) and move it to the Dependent: box. Then select the predictor variables (age and gender) and move them to the Covariates: box. Click on the Categorical button, select gender and move it to the Categorical Covariates box then click Continue > click Save and select Probabilities and Group membership and click Continue > Click Options and select Classification plots, Hosmer-Lemeshow goodness-of-fit and CI for exp(B) then click the Continue then OK as shown in Figure 5.2.

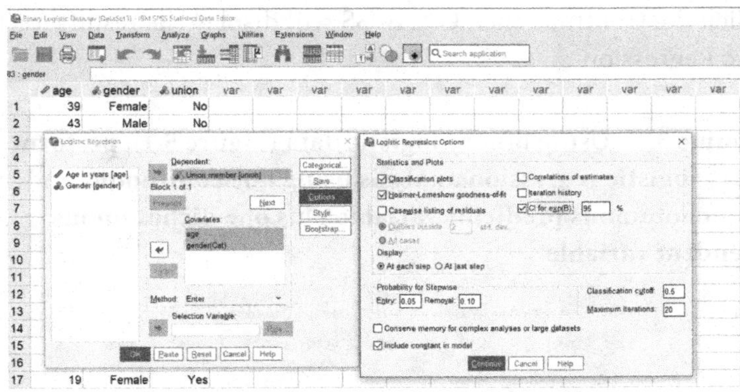

FIGURE 5.2 SPSS Procedure for Binary Logistic Regression

SPSS will display the results for Binary Logistic Regression for the model summary (Figure 5.3) and classification table plus variables in the equation (Figure 5.4).

Model Summary

Step	-2 Log likelihood	Cox & Snell R Square	Nagelkerke R Square
1	62.818[a]	.015	.028

a. Estimation terminated at iteration number 5 because parameter estimates changed by less than .001.

Hosmer and Lemeshow Test

Step	Chi-square	df	Sig.
1	8.212	8	.413

FIGURE 5.3 SPSS Output for Logistic Regression Model Summary

Classification Table[a]

Observed			Predicted: Union member No	Predicted: Union member Yes	Percentage Correct
Step 1	Union member	No	69	0	100.0
		Yes	11	0	.0
	Overall Percentage				86.3

a. The cut value is .500

Variables in the Equation

		B	S.E.	Wald	df	Sig.	Exp(B)	95% C.I.for EXP(B) Lower	95% C.I.for EXP(B) Upper
Step 1[a]	Age in years	.016	.018	.764	1	.382	1.016	.980	1.054
	Gender(1)	-.385	.658	.343	1	.558	.680	.187	2.472
	Constant	-2.443	1.082	5.093	1	.024	.087		

a. Variable(s) entered on step 1: Age in years, Gender.

FIGURE 5.4 SPSS Output for Logistic Regression Classification Table and Variables

5.2.1.3 Multinomial Logistic Regression

Multinomial logistic regression is appropriate when the outcome is a polytomous variable (i.e. categorical with more than two categories), and the predictors are of any type: nominal, ordinal and/or interval/ratio (numeric). Multinomial logistic regression does not require the use of a coding strategy (i.e. dummy coding, effects coding, etc.) for including categorical predictors in the model. Categorical predictor variables can be included directly as factors in the multinomial logistic regression dialog menu box.

Multinomial logistic regression to predict membership of more than two categories. It (basically) works in the same way as binary logistic regression. The analysis breaks the outcome variable down into a series of comparisons between two categories.

For example, if you have three outcome categories (A, B and C), then the analysis will consist of two comparisons that you choose:

- Compare everything against your first category (e.g. A vs. B and A vs. C).
- Or your last category (e.g. A vs. C and B vs. C).
- Or a custom category (e.g. B vs. A and B vs. C).

The general multinomial logistic regression model for each category j (for $j = 1, 2, \ldots, K-1$) is given by

$$P(Y = j|X) = \frac{\exp\left(\beta_{j0} + \sum_{k=1}^{p} \beta_{jk} X_k\right)}{1 + \sum_{m=1}^{K-1} exp\left(\beta_{m0} + \sum_{k=1}^{p} \beta_{mk} X_k\right)}.$$

and for the K reference category

$$P(Y = K|X) = \frac{1}{1 + \sum_{m=1}^{K-1} \exp\left(\beta_{m0} + \sum_{k=1}^{p} \beta_{mk} X_k\right)}.$$

where:

$P(Y = j|X)$ is the probability that the outcome is category j given the predictors X.

$P(Y = K|X)$ is the probability of the reference category given the predictors X.

β_{j0} is the intercept term for category j.
$\beta_{j1}, \beta_{j2}, \ldots, \beta_{jp}$ are the coefficients for the predictors for category j.

To perform Multinomial Logistic Regression in SPSS, we go to: Analyze > Regression > Multinomial Logistic > select the outcome variable (categorical) and move it to the Dependent: box. > click Reference Category and select First Category then click the Continue button > select the predictor variables (numeric) and move them to the Covariate(s): box. If we had any categorical predictors, we would move them to the Factor(s): box > click Statistics and select case processing summary, Pseudo R-square, Step summary, Model fitting information, Information criteria, Classification table, Goodness-of-fit, Estimates, Confidence Interval (%) and Likelihood ratio tests then click the Continue then OK. SPSS Output will display the results of multinomial logistic regression.

Example 5.2: Using multinomial regression data (Figure 5.5). Perform multinomial regression analysis.

Multinomial Logistic Data.sav [DataSet4] - IBM SPSS Statistics Data Editor

File Edit View Data Transform Analyze Graphs Utilities Extensio

32 : edcat

	age	ed	edcat	inccat	var
1	39	17	College ...	\$25 - \$49	
2	43	16	Some co...	\$25 - \$49	
3	49	14	High sch...	\$25 - \$49	
4	35	9	Did not c...	Under \$25	
5	32	16	Some co...	\$25 - \$49	
6	21	14	High sch...	\$25 - \$49	
7	79	12	High sch...	Under \$25	
8	64	11	Did not c...	\$50 - \$74	
9	62	16	Some co...	Under \$25	
10	61	9	Did not c...	\$75 - \$124	
11	48	17	College ...	\$75 - \$124	
12	35	16	Some co...	\$25 - \$49	
13	25	18	College ...	\$25 - \$49	
14	61	16	Some co...	\$75 - \$124	
15	57	16	Some co...	\$125+	
16	52	18	College ...	\$50 - \$74	
17	19	14	High sch...	\$25 - \$49	
18	65	14	High sch...	\$75 - \$124	
19	22	17	College ...	Under \$25	

FIGURE 5.5 Multinomial Regression Data

We go to: Analyze > Regression > Multinomial Logistic > select the outcome variable (inccat) and move it to the Dependent: box. > click Reference Category and select First Category then click the Continue button > select the continuous predictor variables (age and ed) and move them to the Covariate(s): box. Select the categorical predictors (edcat) and move them to the Factor(s): box > click Statistics and select case processing summary, Pseudo R-square, Step summary, Model fitting information, Information criteria, Classification table, Goodness-of-fit, Estimates, Confidence Interval (%) and Likelihood ratio tests then click the Continue then OK (Figure 5.6).

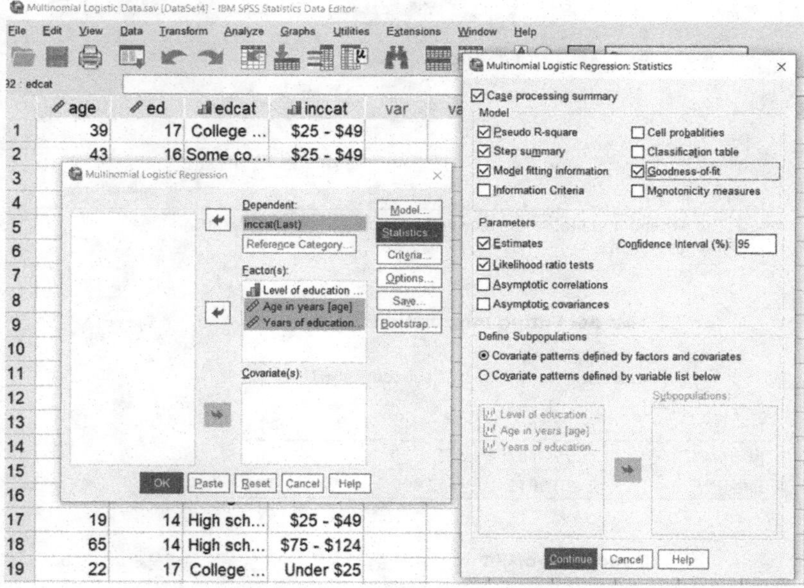

FIGURE 5.6 SPSS Procedure for Multinomial Logistic Regression Analysis

SPSS Output will display the results of multinomial logistic regression (Figures 5.7–5.9).

Case Processing Summary

		N	Marginal Percentage
Income category in thousands	Under $25	20	25.0%
	$25 - $49	32	40.0%
	$50 - $74	11	13.8%
	$75 - $124	11	13.8%
	$125+	6	7.5%
Level of education	Did not complete high school	15	18.8%
	High school degree	33	41.3%
	Some college	17	21.3%
	College degree	13	16.3%
	Post-undergraduate degree	2	2.5%
Valid		80	100.0%
Missing		0	
Total		80	
Subpopulation		76[a]	

a. The dependent variable has only one value observed in 74 (97.4%) subpopulations.

Model Fitting Information

Model	Model Fitting Criteria	Likelihood Ratio Tests		
	-2 Log Likelihood	Chi-Square	df	Sig.
Intercept Only	229.707			
Final	187.642	42.064	24	.013

Goodness-of-Fit

	Chi-Square	df	Sig.
Pearson	242.212	276	.930
Deviance	184.870	276	1.000

FIGURE 5.7 SPSS Output Displaying Case Processing Summary and Model Fitting Information for Multinomial Logistic Regression

Pseudo R-Square

Cox and Snell	.409
Nagelkerke	.433
McFadden	.181

Likelihood Ratio Tests

Effect	Model Fitting Criteria -2 Log Likelihood of Reduced Model	Likelihood Ratio Tests Chi-Square	df	Sig.
Intercept	187.642[a]	.000	0	.
Age in years	199.520	11.878	4	.018
Years of education	189.949	2.306	4	.680
Level of education	201.658	14.016	16	.598

The chi-square statistic is the difference in -2 log-likelihoods between the final model and a reduced model. The reduced model is formed by omitting an effect from the final model. The null hypothesis is that all parameters of that effect are 0.

a. This reduced model is equivalent to the final model because omitting the effect does not increase the degrees of freedom.

FIGURE 5.8 SPSS Output Displaying Pseudo R-Square and Likelihood Ratio Tests for Multinomial Logistic Regression

Parameter Estimates

Income category in thousands[a]		B	Std. Error	Wald	df	Sig.	Exp(B)	95% Confidence Interval for Exp (B) Lower Bound	Upper Bound
Under $25	Intercept	33.048	21.920	2.273	1	.132			
	Age in years	-.068	.043	2.498	1	.114	.934	.858	1.017
	Years of education	-1.437	1.181	1.481	1	.224	.238	.023	2.404
	[Level of education=1]	-.688	479.258	.000	1	.999	.503	.000	.[b]
	[Level of education=2]	4.195	360.657	.000	1	.991	66.359	6.768E-306	.[b]
	[Level of education=3]	-5.769	3.605	2.561	1	.110	.003	2.669E-6	3.656
	[Level of education=4]	-3.568	.000	.	1	.	.028	.028	.028
	[Level of education=5]	0[c]	.	.	0	.	.	.	.
$25 - $49	Intercept	35.385	8634.041	.000	1	.997			
	Age in years	-.087	.043	4.138	1	.042	.916	.843	.997
	Years of education	-1.488	1.174	1.606	1	.205	.226	.023	2.256
	[Level of education=1]	-.770	8647.304	.000	1	1.000	.463	.000	.[b]
	[Level of education=2]	3.473	8641.543	.000	1	1.000	32.233	.000	.[b]
	[Level of education=3]	-5.401	8634.014	.000	1	1.000	.005	.000	.[b]
	[Level of education=4]	-3.999	8634.013	.000	1	1.000	.018	.000	.[b]
	[Level of education=5]	0[c]	.	.	0	.	.	.	.
$50 - $74	Intercept	30.360	22.403	1.837	1	.175			
	Age in years	-.044	.045	.948	1	.330	.957	.876	1.046
	Years of education	-1.387	1.206	1.325	1	.250	.250	.024	2.652
	[Level of education=1]	.207	479.263	.000	1	1.000	1.230	.000	.[b]
	[Level of education=2]	4.449	360.660	.000	1	.990	85.527	8.672E-306	.[b]
	[Level of education=3]	-6.308	3.830	2.713	1	.100	.002	1.002E-6	3.313
	[Level of education=4]	-3.877	.000	.	1	.	.021	.021	.021
	[Level of education=5]	0[c]	.	.	0	.	.	.	.
$75 - $124	Intercept	52.219	5904.484	.000	1	.993			
	Age in years	-.029	.046	.403	1	.526	.971	.888	1.063
	Years of education	-1.551	1.216	1.625	1	.202	.212	.020	2.301
	[Level of education=1]	-21.369	5923.859	.000	1	.997	5.245E-10	.000	.[b]
	[Level of education=2]	-16.801	5915.444	.000	1	.998	5.052E-8	.000	.[b]
	[Level of education=3]	-25.425	5904.441	.000	1	.997	9.080E-12	.000	.[b]
	[Level of education=4]	-23.465	5904.439	.000	1	.997	6.448E-11	.000	.[b]
	[Level of education=5]	0[c]	.	.	0	.	.	.	.

a. The reference category is: $125+.

b. Floating point overflow occurred while computing this statistic. Its value is therefore set to system missing.

c. This parameter is set to zero because it is redundant.

FIGURE 5.9 SPSS Output Displaying Parameter Estimates for Multinomial Logistic Regression

5.2.1.4 Ordinal Logistic Regression

This is used to fit an ordinal dependent (response) variable on a number of predictors (which can be factors or covariates). In general, the ordinal variable has more than two levels. For example, a variable that can take the values low, medium or high.

In ordinal logistic regression, we consider cumulative probability. Instead of considering the probability of an individual event, you consider the probabilities of that event and all events that are ordered before it.

The general ordinal logistic regression model can be written as

$$log\left(\frac{P(Y \leq j \mid X)}{P(Y > j \mid X)}\right) = \alpha_j + \sum_{k=1}^{p} \beta_k X_k \quad for \quad j = 1, 2, 3, \ldots, K-1.$$

where:

α_j is the category-specific intercepts (thresholds).

β_k are the coefficients for the predictor variable's X_k which are the same across all categories.

$P(Y \leq j \mid X)$ is the cumulative probability of the response being in category j or below given the predictors X.

$P(Y > j \mid X)$ is the probability of response being in a category higher than j.

$\beta_1, \beta_2, \ldots, \beta_p$ are the coefficients for the predictors, assumed to be the same for all categories j.

Five different link functions are available in the ordinal regression procedure in SPSS: logit, complementary log-log, negative log-log, probit and Cauchit (inverse Cauchy).

To perform Ordinal Logistic Regression in SPSS, we go to: Analyze > Regression > Ordinal > Transfer the ordinal dependent variable into the Dependent: box. Transfer the categorical independent variables into the Factor(s) box (if more than two levels, otherwise move to covariates box) and the continuous independent variable into the Covariate(s) box > Click Options and select Logit under Link then click Continue > Click Output and select Goodness of fit statistics, Summary statistics, Parameter estimates, Test of parallel lines, Estimated response probabilities, Predicted category, Predicted category probability and Actual category probability then click Continue > Click Location.

Example 5.3: Using ordinal regression data (Figure 5.10). Perform ordinal regression analysis.

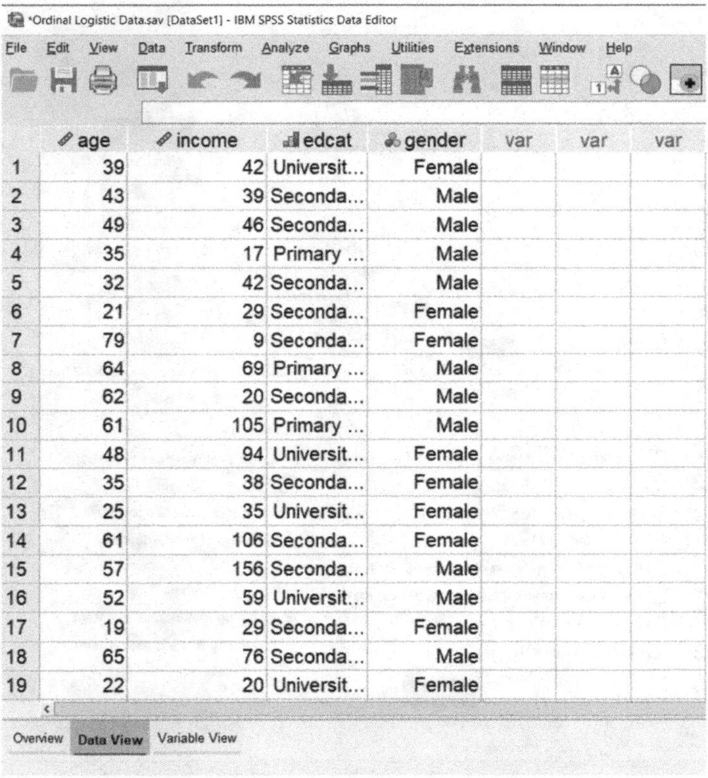

*Ordinal Logistic Data.sav [DataSet1] - IBM SPSS Statistics Data Editor

File Edit View Data Transform Analyze Graphs Utilities Extensions Window Help

	age	income	edcat	gender	var	var	var
1	39	42	Universit...	Female			
2	43	39	Seconda...	Male			
3	49	46	Seconda...	Male			
4	35	17	Primary ...	Male			
5	32	42	Seconda...	Male			
6	21	29	Seconda...	Female			
7	79	9	Seconda...	Female			
8	64	69	Primary ...	Male			
9	62	20	Seconda...	Male			
10	61	105	Primary ...	Male			
11	48	94	Universit...	Female			
12	35	38	Seconda...	Female			
13	25	35	Universit...	Female			
14	61	106	Seconda...	Female			
15	57	156	Seconda...	Male			
16	52	59	Universit...	Male			
17	19	29	Seconda...	Female			
18	65	76	Seconda...	Male			
19	22	20	Universit...	Female			

Overview Data View Variable View

FIGURE 5.10 Ordinal Logistic Data

We go to: Analyze > Regression > Ordinal > Transfer the ordinal dependent variable (level of education) into the Dependent: box. Transfer the categorical independent variables into the Factor(s) box (if more than two levels, otherwise move to covariates box) and the continuous independent variable into the Covariate(s) box. We select age, household income and gender then move them to Covariate(s) box > Click Options and select Logit under Link then click Continue > Click Output and select Goodness of fit statistics, Summary statistics, Parameter estimates, Test of parallel lines, Estimated response probabilities, Predicted category, Predicted category probability and Actual category probability then click Continue then click OK (Figure 5.11).

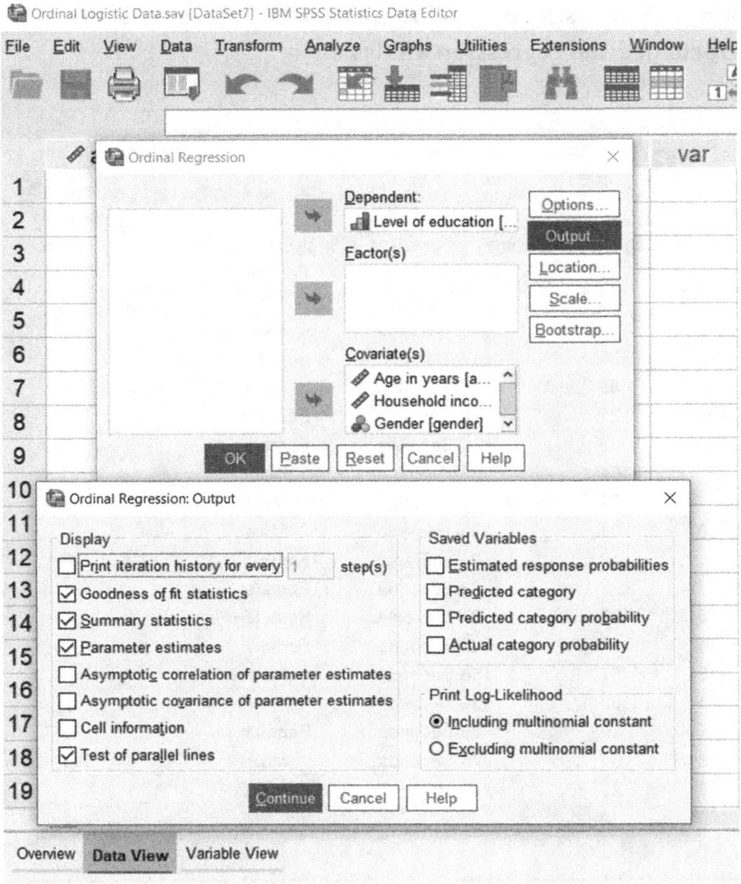

FIGURE 5.11 SPSS Procedure for Ordinal Logistic Regression Analysis

SPSS will display the outputs for ordinal logistic regression analysis (Figure 5.12).

Case Processing Summary

		N	Marginal Percentage
Level of education	Primary education	17	21.3%
	Secondary education	50	62.5%
	University education	13	16.3%
Valid		80	100.0%
Missing		0	
Total		80	

FIGURE 5.12 SPSS Output Displaying Case Summary for Ordinal Logistic Regression Analysis

The case processing summary (Figure 5.12) tells you the proportion of cases falling at each level of the dependent variable (level of education).

Model Fitting Information

Model	-2 Log Likelihood	Chi-Square	df	Sig.
Intercept Only	144.131			
Final	134.260	9.872	3	.020

Link function: Logit.

Goodness-of-Fit

	Chi-Square	df	Sig.
Pearson	142.335	151	.681
Deviance	131.487	151	.872

Link function: Logit.

Pseudo R-Square

Cox and Snell	.116
Nagelkerke	.138
McFadden	.067

Link function: Logit.

FIGURE 5.13 SPSS Output Displaying Model Fitting Information for Ordinal Logistic Regression

The model fitting information (Figure 5.13) contains the –2 log likelihood for an intercept only (or null) model and the full model (has full sets of predictors). The likelihood ratio chi-square test tests whether there is a significant improvement in the fit of the final model relative to the intercept only model. There was a significant improvement in fit for the final model over the null model.

The Goodness of Fit table has the deviance and Pearson chi-square tests that are useful for determining whether a model exhibits good fit for the data. Non-significant test results indicates that the model fits the data well. The analysis results show that both Pearson and deviance test were both nonsignificant hence good model fit.

The pseudo-R-square value is the same as the R square value in OLS regression.

Parameter Estimates

		Estimate	Std. Error	Wald	df	Sig.	95% Confidence Interval Lower Bound	Upper Bound
Threshold	[edcat = 1]	-1.154	.755	2.336	1	.126	-2.633	.326
	[edcat = 2]	2.098	.787	7.111	1	.008	.556	3.641
Location	age	-.019	.013	2.022	1	.155	-.045	.007
	income	.014	.006	6.152	1	.013	.003	.025
	gender	.855	.482	3.142	1	.076	-.090	1.800

Link function: Logit.

FIGURE 5.14 SPSS Output Displaying Regression Coefficients

We have regression coefficients and significance for each of the independent variables (Figure 5.14).

Test of Parallel Lines[a]

Model	-2 Log Likelihood	Chi-Square	df	Sig.
Null Hypothesis	134.260			
General	129.620	4.639	3	.200

The null hypothesis states that the location parameters (slope coefficients) are the same across response categories.

a. Link function: Logit.

FIGURE 5.15 SPSS Output Displaying Test of Parallel Lines

We also have the results for test of Parallel lines (i.e. assumption for proportional odds) (Figure 5.15). The nonsignificance of the test results indicates that the assumption has been satisfied. From the analysis, the assumptions of proportional odds were satisfied.

In SPSS, we can also use the path: Analyze > Generalized Linear Model > Generalized Linear Model > click Type of Model and select ordinal Logistic > click on the Response and select the dependent variable of interest (level of education) > click on predictors and select the independent variables and select age, gender and income as Covariates > click Model tab and specify the model effects and the main effects > click Estimation tab and select the estimation methods > click Statistics tab and select the appropriate statistics and parameters then click OK, and SPSS will generate output for ordinal logistic regression.

5.2.1.5 Probit Models

The probit function is the inverse of the cumulative distribution function of the standard normal distribution to the response proportions, denoted by $\phi(Z)$.

The procedure measures the relationship between the strength of a stimulus and the proportion of cases exhibiting a certain response to the stimulus. It is useful for situations where you have a dichotomous output that is thought to be influenced or caused by levels of some independent variable(s) and is particularly well suited to experimental data. This procedure will allow you to estimate the strength of a stimulus required to induce a certain proportion of responses, such as the median effective dose.

Probit analysis is used to model dichotomous or binary dependent variables. The general expression for probit model is given by

$$\text{Probit}\left[P(X)\right]=\alpha+\beta X, P(X)=\phi(\alpha+\beta X).$$

where:

$\Phi(.)$ is the CDF of the standard normal distribution.
$\beta_1, \beta_2, \ldots, \beta_p$ are the coefficients for predictor variables.
α is the intercept term.

The parameters are estimate using maximum likelihood estimation (MLE) that involves finding the values of β that maximizes the log-likelihood function.

Probit and logit models are reasonable choices when the changes in the cumulative probabilities are gradual. In practice, probit and logistic regression models provide similar fits.

If a logistic regression model fits well, then so does the probit model, and conversely.

In general, probit analysis is appropriate for designed experiments, whereas logistic regression is more appropriate for observational studies.

To perform Probit Regression in SPSS, we go to: Analyze > Regression > Probit > Select a response frequency variable. This variable indicates the number of cases exhibiting a response to the test stimulus > Select a total observed variable. This variable indicates the number of cases to which the stimulus was applied > Select one or more covariate(s). This variable contains the level of the stimulus applied to each observation > select the Probit Model.

We can also use the path: Analyze > Generalized Linear Models > Generalized Linear Models > under Type of Model select Binary probit > click Response and move categorical response variable into the Dependent variable box then click Reference Category and select those in union i.e.

code = 1 or Last value > click Predictors and move all the predictors to the Covariates box. If you have categorical variable with more than two levels, move them to Factors box > click Model and move all the predictor variables into the Model box > click Statistics and add to the defaults include exponential parameter estimates then click OK.

Example 5.4: Using probit data (Figure 5.16), perform probit regression analysis to predict whether a given person is in union or not using age, income and gender.

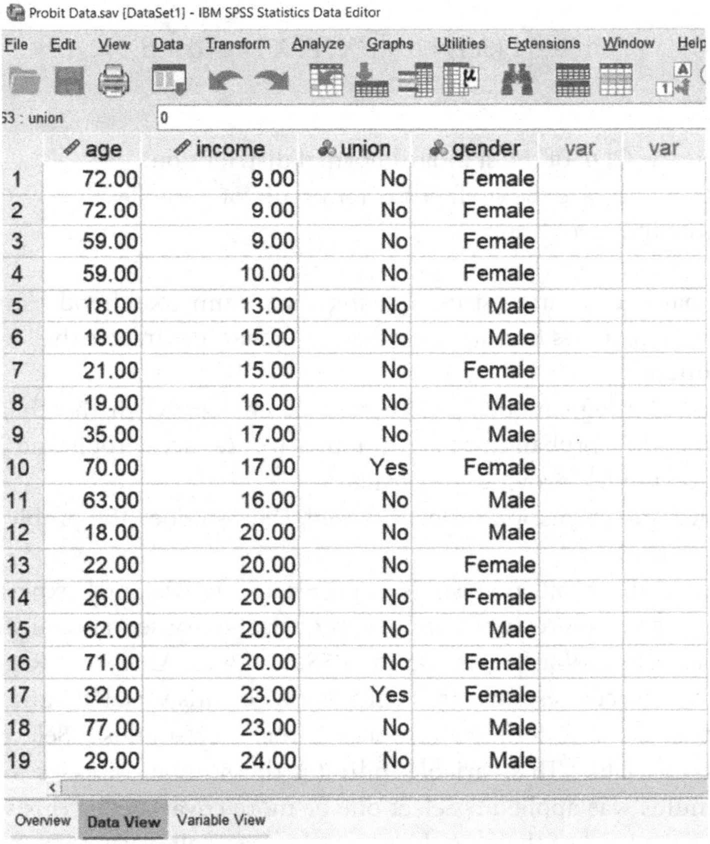

Probit Data.sav [DataSet1] - IBM SPSS Statistics Data Editor

File Edit View Data Transform Analyze Graphs Utilities Extensions Window Help

53 : union 0

	age	income	union	gender	var	var
1	72.00	9.00	No	Female		
2	72.00	9.00	No	Female		
3	59.00	9.00	No	Female		
4	59.00	10.00	No	Female		
5	18.00	13.00	No	Male		
6	18.00	15.00	No	Male		
7	21.00	15.00	No	Female		
8	19.00	16.00	No	Male		
9	35.00	17.00	No	Male		
10	70.00	17.00	Yes	Female		
11	63.00	16.00	No	Male		
12	18.00	20.00	No	Male		
13	22.00	20.00	No	Female		
14	26.00	20.00	No	Female		
15	62.00	20.00	No	Male		
16	71.00	20.00	No	Female		
17	32.00	23.00	Yes	Female		
18	77.00	23.00	No	Male		
19	29.00	24.00	No	Male		

Overview Data View Variable View

FIGURE 5.16 Prohibit Data

Analyze > Generalized Linear Models > Generalized Linear Models > under Type of Model select Binary probit > click Response and move union into the Dependent variable box then click Reference Category and select those in union i.e. code = 1 or Last value > click Predictors and move all the predictors to the Covariates box. If you have categorical variable with more than two levels, move them to Factors box > click Model and move all the predictor variables into the Model box > click Statistics and add to the defaults include exponential parameter estimates then click OK (Figure 5.17).

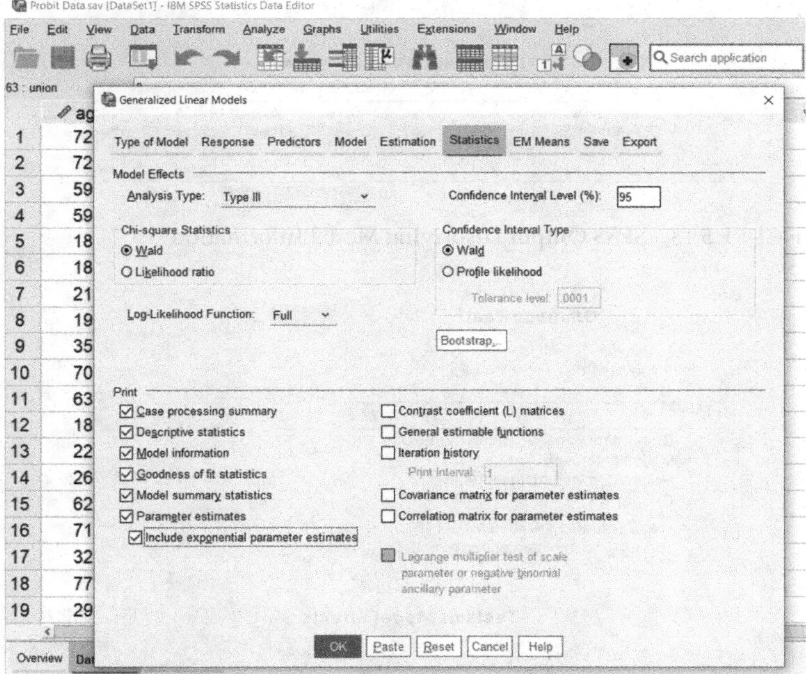

FIGURE 5.17 SPSS Procedure for Probit Regression Analysis

SPSS will display the outputs for probit regression analysis.

Model Information

Dependent Variable	Union member[a]
Probability Distribution	Binomial
Link Function	Probit

a. The procedure models No as the response, treating Yes as the reference category.

Case Processing Summary

	N	Percent
Included	80	100.0%
Excluded	0	0.0%
Total	80	100.0%

Categorical Variable Information

			N	Percent
Dependent Variable	Union member	No	69	86.3%
		Yes	11	13.8%
		Total	80	100.0%

Continuous Variable Information

		N	Minimum	Maximum	Mean	Std. Deviation
Covariate	Age in years	80	18.00	78.00	46.9125	17.24521
	Household income in thousands	80	9.00	146.00	47.6375	34.36964
	Gender	80	.00	1.00	.5250	.50253

FIGURE 5.18 SPSS Output Displaying Model Information

Omnibus Test[a]

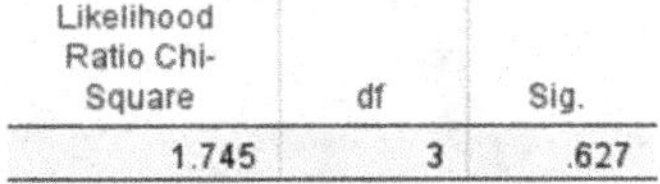

Likelihood Ratio Chi-Square	df	Sig.
1.745	3	.627

Dependent Variable: Union member
Model: (Intercept), Age in years, Household income in thousands, Gender

a. Compares the fitted model against the intercept-only model.

Tests of Model Effects

	Type III		
Source	Wald Chi-Square	df	Sig.
(Intercept)	1.813	1	.178
Age in years	.187	1	.665
Household income in thousands	.574	1	.449
Gender	.423	1	.516

Dependent Variable: Union member
Model: (Intercept), Age in years, Household income in thousands, Gender

FIGURE 5.19 SPSS Output Displaying Omnibus Test

Parameter Estimates										
			95% Wald Confidence Interval		Hypothesis Test				95% Wald Confidence Interval for Exp(B)	
Parameter	B	Std. Error	Lower	Upper	Wald Chi-Square	df	Sig.	Exp(B)	Lower	Upper
(Intercept)	.785	.5829	-.358	1.927	1.813	1	.178	2.192	.699	6.870
Age in years	.005	.0109	-.017	.026	.187	1	.665	1.005	.984	1.026
Household income in thousands	.005	.0067	-.008	.018	.574	1	.449	1.005	.992	1.018
Gender	-.236	.3627	-.947	.475	.423	1	.516	.790	.388	1.608
(Scale)	1[a]									

Dependent Variable: Union member
Model: (Intercept), Age in years, Household income in thousands, Gender
a. Fixed at the displayed value.

FIGURE 5.20 SPSS Output Displaying Parameter Estimates

The findings shows model information (Figure 5.18), the omnibus test (Figure 5.19) and the parameter estimates table (Figure 5.20).

5.3 MODELS FOR COUNT

5.3.1 Count Data

Count data are common in many disciplines. A common example is when the response variable is the counted number of occurrences of an event. The distribution of counts is discrete, not continuous and is limited to non-negative values. Count models can be used for rate data in many instances by using exposure.

Regression models with count data include Poisson regression, negative binomial regression, zero-inflated count models (zero-inflated Poisson and zero-inflated negative binomial), zero-truncated count models (zero-truncated Poisson, zero-truncated negative binomial), hurdle models and random-effect count models.

Negative binomial distribution can be used to model count data with over-dispersion.

Zero-inflated models attempt to account for excess zeros i.e. there is thought to be two kinds of zeros: "true zeros" and excess zeros. Zero-inflated models estimate two equations: one for the count model and one for the excess zeros.

Zero-truncated models used when the count variable cannot take on the value zero.

A hurdle model is a modified count model in which there are two processes: one generating the zeros and one generating the positive values. The two models are not constrained to be the same. The concept underlying the hurdle model is that a binomial probability model governs the binary outcome of whether a count variable has a zero or a positive value. If the value is positive, the "hurdle is crossed", and the conditional distribution of the positive values is governed by a zero-truncated count model.

5.3.2 Poisson Regression

Poisson distribution function of the (count) variable y is given by

$$f(y) = P(Y = y) = \frac{e^{-\mu}\mu^{y}}{y!}.$$

where μ = the rate, y = successes.

The general poison regression model with a single predictor is given y.

$$E(y) = \mu = \exp(\beta_0 + \beta_1 \mathbf{x}).$$

where:

β_0 is the intercept.

β_1 is the coefficient of the predictor variable.

Poisson regression is used to predict a dependent variable that consists of "count data" given one or more independent variables.

The Poisson model assumes that the conditional mean and variance of the outcome are approximately equal (i.e. mean and variance taking into account the covariates in the model).

To perform Poison Regression in SPSS, we go to: Click Analyze > Generalized Linear Models > Generalized Linear Models > select Poisson loglinear in the Counts area > Select the Response tab and transfer your dependent variable (ordinal data) into the Dependent variable: box > select the Predictors tab and transfer the categorical independent variable into the Factors: box and the continuous independent variable into the Covariates: box > select the Model tab and keep the default of Main effects in the – Build Term(s) – area then transfer the categorical and continuous independent variables from the Factors and Covariates: box into the Model: box > select the Estimation tab and keep the default options selected > select the Statistics tab and keep the defaults and select Include exponential parameter estimates in the Print area then click OK. SPSS Output will give the results for the Poisson Regression.

Example 5.5: Using Poisson data (Figure 5.21), perform Poisson regression analysis to determine if the number of cars owned depends on type of gender and income.

Poisson Data.sav [DataSet1] - IBM SPSS Statistics Data Editor

File Edit View Data Transform Analyze Graphs Utilities Extensio

95 : gender

	income	gender	cars	var
1	9.00	Female	1	
2	9.00	Female	4	
3	9.00	Female	1	
4	10.00	Female	3	
5	13.00	Male	3	
6	15.00	Male	1	
7	15.00	Female	4	
8	16.00	Male	1	
9	17.00	Male	3	
10	17.00	Female	1	
11	16.00	Male	2	
12	20.00	Male	0	
13	20.00	Female	2	
14	20.00	Female	2	
15	20.00	Male	1	
16	20.00	Female	1	
17	23.00	Female	5	
18	23.00	Male	2	
19	24.00	Male	0	

Overview Data View Variable View

FIGURE 5.21 Poisson Data

We go to: Click Analyze > Generalized Linear Models > Generalized Linear Models > select Poisson loglinear in the Counts area > Select the Response tab and transfer your dependent variable (ordinal data – cars) into the Dependent variable: box > select the Predictors tab and transfer the categorical independent variable (gender) into the Factors: box and the continuous independent variable (income) into the Covariates: box > select the Model tab and keep the default of Main effects in the – Build Term(s) – area then transfer the categorical and continuous independent variables (income and gender) from the Factors and Covariates: box into the Model: box > select the Estimation tab and keep the default options selected > select the Statistics tab and keep the defaults and select Include exponential parameter estimates in the Print area then click OK (Figure 5.22).

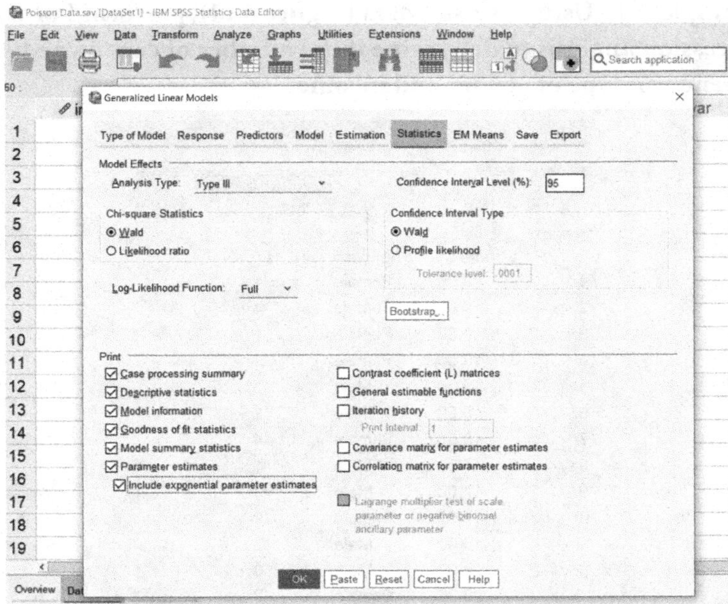

FIGURE 5.22 SPSS procedure for Poisson Regression Analysis

SPSS Output will give the results for the Poisson Regression (Figures 5.23 and 5.24).

Model Information

Dependent Variable	Number of cars owned/leased
Probability Distribution	Poisson
Link Function	Log

Case Processing Summary

	N	Percent
Included	80	100.0%
Excluded	0	0.0%
Total	80	100.0%

Categorical Variable Information

			N	Percent
Factor	Gender	Male	38	47.5%
		Female	42	52.5%
		Total	80	100.0%

Continuous Variable Information

		N	Minimum	Maximum	Mean	Std. Deviation
Dependent Variable	Number of cars owned/leased	80	0	6	1.93	1.367
Covariate	Household income in thousands	80	9.00	146.00	47.6375	34.36964

FIGURE 5.23 SPSS Output Displaying Model Information, Case Processing Summary and Variable Information for Poisson Regression

Goodness of Fit[a]

	Value	df	Value/df
Deviance	84.395	77	1.096
Scaled Deviance	84.395	77	
Pearson Chi-Square	72.930	77	.947
Scaled Pearson Chi-Square	72.930	77	
Log Likelihood[b]	-131.771		
Akaike's Information Criterion (AIC)	269.542		
Finite Sample Corrected AIC (AICC)	269.858		
Bayesian Information Criterion (BIC)	276.688		
Consistent AIC (CAIC)	279.688		

Dependent Variable: Number of cars owned/leased
Model: (Intercept), Gender, Household income in thousands

a. Information criteria are in smaller-is-better form.

b. The full log likelihood function is displayed and used in computing information criteria.

Omnibus Test[a]

Likelihood Ratio Chi-Square	df	Sig.
3.218	2	.200

Dependent Variable: Number of cars owned/leased
Model: (Intercept), Gender, Household income in thousands

a. Compares the fitted model against the intercept-only model.

FIGURE 5.24 SPSS Output Displaying Goodness of Fit and Omnibus Test for Poisson Regression

The values of Pearson chi-square i.e. value/df will be used to determine if there is equi-dispersion (χ^2=1), over dispersion (χ^2>1) and under dispersion (χ^2<1). From Figure 5.24 we have overdispersion χ^2=3.218.

Tests of Model Effects

	Type III		
Source	Wald Chi-Square	df	Sig.
(Intercept)	16.203	1	<.001
Gender	2.829	1	.093
Household income in thousands	.531	1	.466

Dependent Variable: Number of cars owned/leased
Model: (Intercept), Gender, Household income in thousands

Parameter Estimates

			95% Wald Confidence Interval		Hypothesis Test				95% Wald Confidence Interval for Exp(B)	
Parameter	B	Std. Error	Lower	Upper	Wald Chi-Square	df	Sig.	Exp(B)	Lower	Upper
(Intercept)	.697	.1493	.405	.990	21.813	1	<.001	2.009	1.499	2.692
[Gender=.00]	-.276	.1643	-.598	.046	2.829	1	.093	.759	.550	1.047
[Gender=1.00]	0[a]	.	.	.	.	.	.	1	.	.
Household income in thousands	.002	.0023	-.003	.006	.531	1	.466	1.002	.997	1.006
(Scale)	1[b]									

Dependent Variable: Number of cars owned/leased
Model: (Intercept), Gender, Household income in thousands

a. Set to zero because this parameter is redundant.

b. Fixed at the displayed value.

FIGURE 5.25 SPSS Output Displaying the Tests for Model Effects and Parameter Estimates for Poisson Regression

5.3.3 Negative Binomial Regression

Negative binomial regression is used to test for associations between predictor and confounding variables on a count outcome variable when the variance of the count is higher than the mean of the count.

When the conditional variance exceeds the conditional mean, which frequently occurs in practice, it is referred to as *overdispersion*. This may bias standard errors and thus statistical tests. The negative binomial model is a related approach but does not require the equal conditional variance and mean, allowing for overdispersion without bias in standard error estimates. When there is no overdispersion, the negative binomial and Poisson are the same.

The general expression for a negative binomial regression with a single predictor is given by

$$\log(\mu) = \beta_0 + \beta_1 X.$$

or

$$\mu = \exp(\beta_0 + \beta_1 X).$$

where:

β_0 is the intercept.
β_1 is the coefficient of the predictor variable.

To perform Negative Binomial Regression in SPSS, we go to: Analyze > Generalized Linear Models > Generalized Linear Model > In the Type of Model tab, under the Customs header select the Negative binomial under Distribution and log under Link Function then click Estimate value > Click Response tab and select the count outcome variable and drag it to the Dependent Variable: box > click the Predictors tab and select the categorical or ordinal predictor variable and move it to the Factors: box. Select a continuous predictor variable and move the variable into the Covariates: box > click the Model tab and keep the default of Main effects in the – Build Term(s) – area then transfer the categorical and continuous independent variables from the Factors and Covariates: box into the Model: box > select the Estimation tab and keep the default options selected > select the Statistics tab and keep the defaults and select Include exponential parameter estimates in the Print area then click OK. SPSS Output will have the results for the Negative Binomial Regression.

Example 5.6: Using the Poisson data, perform negative binomial regression.

Poisson Data.sav [DataSet1] - IBM SPSS Statistics Data Editor

File Edit View Data Transform Analyze Graphs Utilities Extensio

95 : gender

	income	gender	cars	var
1	9.00	Female	1	
2	9.00	Female	4	
3	9.00	Female	1	
4	10.00	Female	3	
5	13.00	Male	3	
6	15.00	Male	1	
7	15.00	Female	4	
8	16.00	Male	1	
9	17.00	Male	3	
10	17.00	Female	1	
11	16.00	Male	2	
12	20.00	Male	0	
13	20.00	Female	2	
14	20.00	Female	2	
15	20.00	Male	1	
16	20.00	Female	1	
17	23.00	Female	5	
18	23.00	Male	2	
19	24.00	Male	0	

Overview Data View Variable View

FIGURE 5.21 Poisson Data

We first perform Poisson regression analysis to determine if there is overdispersion.

If there is over dispersion from Poison regression (i.e. from the Goodness of fit table for poison regression), we can perform Negative Binomial regression.

If the likelihood ratio chi-square test from omnibus test table is statistically significant it will suggest a better/additional model relative to the normal model.

We go to: Analyze > Generalized Linear Models > Generalized Linear Model > In the Type of Model tab, under the Customs header select the Negative binomial under Distribution and log under Link Function then click Estimate value > Click Response tab and select the count outcome variable (cars) and drag it to the Dependent Variable: box > click the Predictors tab and select the categorical or ordinal predictor variable (gender) and move it to the Factors: box. Select a continuous predictor variable (income) and move the variable into the Covariates: box > click the Model tab and keep the default of Main effects in the – Build Term(s) – area then transfer the categorical and continuous independent variables (income and gender) from the Factors and Covariates: box into the Model: box > select the Estimation tab and keep the default options selected > select the Statistics tab and keep the defaults and select Include exponential parameter estimates in the Print area then click OK (Figure 5.26).

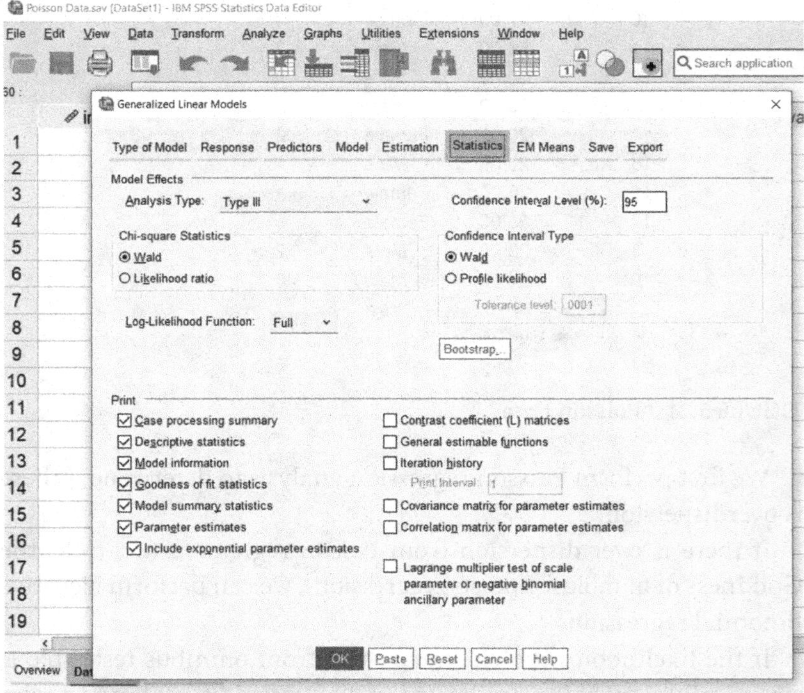

FIGURE 5.26 SPSS Procedure for Negative Binomial Regression Analysis

SPSS Output will have the results for the Negative Binomial Regression.

Model Information

Dependent Variable	Number of cars owned/leased
Probability Distribution	Negative binomial (MLE)
Link Function	Log

Case Processing Summary

	N	Percent
Included	80	100.0%
Excluded	0	0.0%
Total	80	100.0%

Categorical Variable Information

			N	Percent
Factor	Gender	Male	38	47.5%
		Female	42	52.5%
		Total	80	100.0%

Continuous Variable Information

		N	Minimum	Maximum	Mean	Std. Deviation
Dependent Variable	Number of cars owned/leased	80	0	6	1.93	1.367
Covariate	Household income in thousands	80	9.00	146.00	47.6375	34.36964

FIGURE 5.27 SPSS Output Displaying the Model Information for Negative Binomial Regression

Goodness of Fit[a]

	Value	df	Value/df
Deviance	84.395	76	1.110
Scaled Deviance	84.395	76	
Pearson Chi-Square	72.930	76	.960
Scaled Pearson Chi-Square	72.930	76	
Log Likelihood[b]	-131.771		
Akaike's Information Criterion (AIC)	271.542		
Finite Sample Corrected AIC (AICC)	272.075		
Bayesian Information Criterion (BIC)	281.070		
Consistent AIC (CAIC)	285.070		

Dependent Variable: Number of cars owned/leased
Model: (Intercept), Gender, Household income in thousands

a. Information criteria are in smaller-is-better form.

b. The full log likelihood function is displayed and used in computing information criteria.

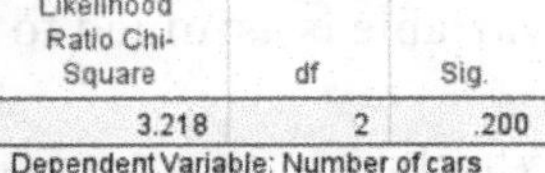

Omnibus Test[a]

Likelihood Ratio Chi-Square	df	Sig.
3.218	2	.200

Dependent Variable: Number of cars owned/leased
Model: (Intercept), Gender, Household income in thousands

a. Compares the fitted model against the intercept-only model.

FIGURE 5.28 SPSS Output Displaying the Goodness of Fit Test for Negative Binomial R

Tests of Model Effects

Source	Type III		
	Wald Chi-Square	df	Sig.
(Intercept)	16.203	1	<.001
Gender	2.829	1	.093
Household income in thousands	.531	1	.466

Dependent Variable: Number of cars owned/leased
Model: (Intercept), Gender, Household income in thousands

Parameter Estimates

Parameter	B	Std. Error	95% Wald Confidence Interval		Hypothesis Test			Exp(B)	95% Wald Confidence Interval for Exp(B)	
			Lower	Upper	Wald Chi-Square	df	Sig.		Lower	Upper
(Intercept)	.697	.1493	.405	.990	21.813	1	<.001	2.009	1.499	2.692
[Gender=.00]	-.276	.1643	-.598	.046	2.829	1	.093	.759	.560	1.047
[Gender=1.00]	0[a]	.	.	.	.	.	.	1	.	.
Household income in thousands	.002	.0023	-.003	.006	.531	1	.466	1.002	.997	1.006
(Scale)	1[b]									
(Negative binomial)	6.933E-8	3.9967E-5	.000							

Dependent Variable: Number of cars owned/leased
Model: (Intercept), Gender, Household income in thousands

a. Set to zero because this parameter is redundant.

b. Fixed at the displayed value.

FIGURE 5.29 SPSS Output Displaying Parameter Estimates for Negative Binomial Regression

The interpretation is the same as that of Poisson Regression. We have the model information (Figure 5.27), goodness of fit (Figure 5.28) where the Pearson chi square is much closer to 1 and parameter estimates (Figure 5.29).

5.4 LOGLINEAR REGRESSION

General loglinear model is a technique for modeling a categorical response variable, which are often count data that follow a Poisson distribution or frequency in a cross-tabulation form that follows a multinomial distribution, based on a set of factors or covariates. The maximum number of factors or covariates allowed is ten. The procedure analyzes the frequency counts of cases falling into each category of a cross classification of two or more variables. Each cross classification leads to a cell. The variables forming the cross classification are called factors. In this analysis, the frequency in each cell is the dependent variable. The dependent variable is assumed to follow a Poisson or multinomial distribution.

Logit loglinear models are similar to ANOVA models for the logit-expected cell frequencies of cross-tabulation tables.

Logit loglinear analysis is used to analyze the relationship between categorical dependent variables and independent variables. The independent variables can be categorical (factors) or scale (covariates). One can have up to ten dependent and factor variables combined. A cell structure variable allows one to define structural zeros or include an offset term in the model.

Loglinear regression analysis is used to describe the pattern of data in a contingency table. A model is constructed to predict the natural log of the frequency of each cell in the contingency table.

The loglinear model specifies that the logarithm of the expected count, $\ddot{e}$, as a linear function of the predictor variables

$$\log(\lambda) = \beta_0 + \beta_1 X_1 + \beta_2 X_2 + \ldots + \beta_p X_p.$$

where:

$\lambda = E(Y|X)$ is the expected count.

β_0 is the intercept term.

$\beta_1, \beta_2, \ldots, \beta_p$ are the coefficients for the predictor variables.

Taking exponents both sides of the loglinear model, we get the mean model:

$$\lambda = \exp\left(\beta_0 + \beta_1 X_1 + \beta_2 X_2 + \ldots + \beta_p X_p\right).$$

The parameters are estimate using maximum likelihood estimation (MLE) that involves finding the values of β that maximizes the log-likelihood function.

In SPSS, we go to: Analyze > Loglinear > Model Selection > Move the categorical variables of interest into the "Factor(s)" box then click Define Range and define the range for the variables by putting code for Minimum and code for the Maximum > click Options and in addition to the defaults select Parameter estimates and Association table then click OK. SPSS Output will have the results for the Loglinear Regression.

Example 5.7: Using loglinear data (Figure 5.30), perform loglinear analysis.

Loglinear Data.sav [DataSet1] - IBM SPSS Statistics Data Editor

File Edit View Data Transform Analyze Graphs Utilities E

28 : union 0

	default	gender	union	var
1	Yes	Female	No	
2	Yes	Female	No	
3	No	Female	No	
4	No	Female	No	
5	No	Male	No	
6	Yes	Male	No	
7	No	Female	No	
8	No	Male	No	
9	No	Male	No	
10	No	Female	Yes	
11	No	Male	No	
12	No	Male	No	
13	Yes	Female	No	
14	No	Female	No	
15	No	Male	No	
16	No	Female	No	
17	No	Female	Yes	
18	No	Male	No	
19	Yes	Male	No	

Overview Data View Variable View

FIGURE 5.30 Loglinear Data

We go to: Analyze > Loglinear > Model Selection > Move the categorical variables of interest (default, gender, union) into the "Factor(s)" box then click Define Range and define the range for the variables by putting 0 on Minimum and 1 on the Maximum > click Options and in addition to the defaults select Parameter estimates and Association table then click OK (Figure 5.31).

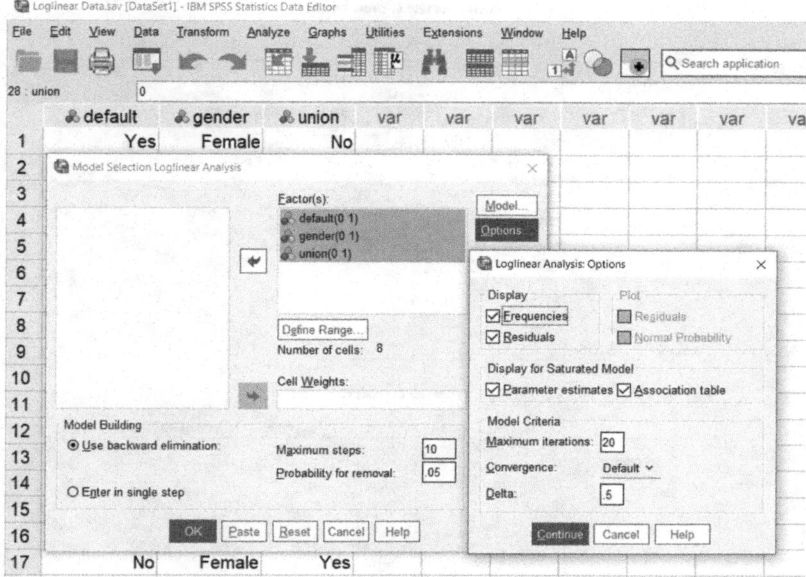

FIGURE 5.31 SPSS procedure for Loglinear Regression Analysis

Data Information

		N
Cases	Valid	80
	Out of Range[a]	0
	Missing	0
	Weighted Valid	80
Categories	Ever defaulted on a bank loan	2
	Gender	2
	Union member	2

a. Cases rejected because of out of range factor values.

Cell Counts and Residuals

Ever defaulted on a bank loan	Gender	Union member	Observed Count[a]	Observed %	Expected Count	Expected %	Residuals	Std. Residuals
No	Male	No	24.500	30.6%	24.500	30.6%	.000	.000
		Yes	3.500	4.4%	3.500	4.4%	.000	.000
	Female	No	25.500	31.9%	25.500	31.9%	.000	.000
		Yes	6.500	8.1%	6.500	8.1%	.000	.000
Yes	Male	No	10.500	13.1%	10.500	13.1%	.000	.000
		Yes	1.500	1.9%	1.500	1.9%	.000	.000
	Female	No	10.500	13.1%	10.500	13.1%	.000	.000
		Yes	1.500	1.9%	1.500	1.9%	.000	.000

a. For saturated models, .500 has been added to all observed cells.

Goodness-of-Fit Tests

	Chi-Square	df	Sig.
Likelihood Ratio	.000	0	.
Pearson	.000	0	.

FIGURE 5.32 SPSS Output Displaying Descriptive Statistics and Goodness-of-Fit Tests for Loglinear Regression

K-Way and Higher-Order Effects

	K	df	Likelihood Ratio Chi-Square	Likelihood Ratio Sig.	Pearson Chi-Square	Pearson Sig.	Number of Iterations
K-way and Higher Order Effects[a]	1	7	65.273	<.001	64.800	<.001	<.001
	2	4	1.437	.838	1.471	.832	2
	3	1	.152	.696	.155	.694	2
K-way Effects[b]	1	3	63.836	<.001	63.329	<.001	<.001
	2	3	1.284	.733	1.316	.725	<.001
	3	1	.152	.696	.155	.694	<.001

a. Tests that k-way and higher order effects are zero.

b. Tests that k-way effects are zero.

Partial Associations

Effect	df	Partial Chi-Square	Sig.	Number of Iterations
default*gender	1	.044	.833	2
default*union	1	.565	.452	2
gender*union	1	.611	.434	2
default	1	16.797	<.001	2
gender	1	.200	.655	2
union	1	46.840	<.001	2

Parameter Estimates

Effect	Parameter	Estimate	Std. Error	Z	Sig.	95% Confidence Interval Lower Bound	95% Confidence Interval Upper Bound
default*gender*union	1	.072	.179	.405	.685	-.278	.423
default*gender	1	-.082	.179	-.461	.645	-.433	.268
default*union	1	-.072	.179	-.405	.685	-.423	.278
gender*union	1	.072	.179	.405	.685	-.278	.423
default	1	.506	.179	2.832	.005	.156	.856
gender	1	-.082	.179	-.461	.645	-.433	.268
union	1	.901	.179	5.040	<.001	.550	1.251

FIGURE 5.33 SPSS Output Displaying K-Way and Higher-Order Effects, Partial Associations and Parameter Estimates for Loglinear Regression

Backward Elimination Statistics

Step Summary

Step[a]			Effects	Chi-Square[c]	df	Sig.	Number of Iterations
0	Generating Class[b]		default*gender*union	.000	0	.	
	Deleted Effect	1	default*gender*union	.152	1	.696	2
1	Generating Class[b]		default*gender, default*union, gender*union	.152	1	.696	
	Deleted Effect	1	default*gender	.044	1	.833	2
		2	default*union	.565	1	.452	2
		3	gender*union	.611	1	.434	2
2	Generating Class[b]		default*union, gender*union	.197	2	.906	
	Deleted Effect	1	default*union	.597	1	.440	2
		2	gender*union	.643	1	.423	2
3	Generating Class[b]		gender*union, default	.794	3	.851	
	Deleted Effect	1	gender*union	.643	1	.423	2
		2	default	16.797	1	<.001	2
4	Generating Class[b]		default, gender, union	1.437	4	.838	
	Deleted Effect	1	default	16.797	1	<.001	2
		2	gender	.200	1	.655	2
		3	union	46.840	1	<.001	2
5	Generating Class[b]		default, union	1.637	5	.897	
	Deleted Effect	1	default	16.797	1	<.001	2
		2	union	46.840	1	<.001	2
6	Generating Class[b]		default, union	1.637	5	.897	

a. At each step, the effect with the largest significance level for the Likelihood Ratio Change is deleted, provided the significance level is larger than .050.

b. Statistics are displayed for the best model at each step after step 0.

c. For 'Deleted Effect', this is the change in the Chi-Square after the effect is deleted from the model.

FIGURE 5.34 SPSS Output Displaying Backward Elimination Statistics for Loglinear Regression

Goodness-of-Fit Tests			
	Chi-Square	df	Sig.
Likelihood Ratio	1.637	5	.897
Pearson	1.704	5	.888

FIGURE 5.35 SPSS Output Displaying Goodness-of-Fit Tests for Loglinear Regression

SPSS Output will have the results for the loglinear regression.

Data Information and Cell and Count Residuals (Figure 5.32) and check if no cell count is less than 5%. In the Goodness of Fit table, it shows a perfect fit model (i.e. no *p*-value). The SPSS output shows Kway and higher effects, Partial associations table and Parameter estimates table for z scores (Figure 5.33), backward elimination table that provide the final results with statistical and significant results (Figure 5.34) and goodness of fit table (Figure 5.35).

5.5 GAMMA

A generalized linear model can be used to fit a Gamma regression for the analysis of positive range data.

A response variable Yfollows a gamma distribution with mean μ and shape parameter k if the probability density function is given by

$$f(y{:}\mu,k)=\frac{y^{k-1}e^{-yk/\mu}}{\left(\frac{\mu}{k}\right)^{k}\Gamma(k)} \quad for \quad y>0.$$

where:

μ is the mean of the distribution.

k is the shape parameter.

$\Gamma(k)$ is the gamma function evaluated at k.

The gamma regression model with a single predictor X is given by

$$\log(\mu)=\exp(\beta_0+\beta_1X).$$

where:

β_0 is the intercept term.

$\beta_1, \beta_2, \ldots, \beta_p$ are the coefficients for the predictor variables.

The parameters $\beta=(\beta_1, \beta_2, \ldots, \beta_p)$ and k are estimated using maximum likelihood estimation (MLE) that involves finding the values of β and k that maximizes the log-likelihood function.

In SPSS, we go to: the Analyze > Generalized Linear Models > Generalized linear models > under Type of Model tab, select Gamma with log link under Scale Response.

We can also use the path: Analyze > Generalized Linear Models > Generalized Linear Model > In the Type of Model tab, under the Customs header, select the Gamma under Distribution and log under Link Function then click Estimate value > Click Response tab and select the count outcome variable and drag it to the Dependent Variable: box > click the Predictors tab and select the categorical or ordinal predictor variable and move it to the Factors: box. Select a continuous predictor variable and move the variable into the Covariates: box > click the Model tab and keep the default of Main effects in the – Build Term(s) – area then transfer the categorical and continuous independent variables from the Factors and Covariates: box into the Model: box > select the Estimation tab and keep the default options selected > select the Statistics tab and keep the defaults and select Include exponential parameter estimates in the Print area then click OK. SPSS Output will have the results for the Gamma Regression.

5.6 INVERSE GAUSSIAN GLMs

Inverse Gaussian generalized linear models (IG-GLMs) are a type of generalized linear model that can be used to model data that are distributed according to the inverse Gaussian distribution. The inverse Gaussian distribution is a continuous distribution that is often used to model data that are positively skewed and have a variance that is proportional to the mean.

A response variable Y follows a gamma distribution with mean μ and shape parameter k if the probability density function is given by

$$f(y;\mu,\lambda) = \left(\frac{\lambda}{2\pi y^3}\right)^{1/2} exp\left(-\frac{\lambda(y-\mu)^2}{2\mu^2 y}\right) \quad for \quad y > 0.$$

where:

μ is the mean of the distribution.
λ is the shape parameter.

The inverse Gaussian GLM with one predictor variable is given by

$$\mu = g^{-1}(\beta_0 + \beta_1 X)$$

where:

$g^{-1}(.)$ is the inverse link function.

β_0 is the intercept term.

β_1 is the coefficients for the predictor variable.

The parameters $\beta = (\beta_1, \beta_2, \ldots, \beta_p)$ and λ are estimated using maximum likelihood estimation (MLE) that involves finding the values of β and λ that maximizes the log-likelihood function.

To perform the Inverse Gaussian Regression in SPSS, we go to: Analyze > Generalized Linear Models > Generalized Linear Model > In the Type of Model tab, under the Customs header, select the Inverse Gaussian under Distribution and log under Link Function then click Estimate value > Click Response tab and select the count outcome variable and drag it to the Dependent Variable: box > click the Predictors tab and select the categorical or ordinal predictor variable and move it to the Factors: box. Select a continuous predictor variable and move the variable into the Covariates: box > click the Model tab and keep the default of Main effects in the – Build Term(s) – area then transfer the categorical and continuous independent variables from the Factors and Covariates: box into the Model: box > select the Estimation tab and keep the default options selected > select the Statistics tab and keep the defaults and select Include exponential parameter estimates in the Print area then click OK. SPSS Output will have the results for the Negative Binomial Regression.

5.7 NON LINEAR REGRESSION

Not all data follow a straight line that it can be fitted using linear models. Some of the relationships that are not often nonlinear include: economic growth rates, disease outbreaks, life expectancy, fuel consumption, power output, productivity gains etc. Also, from the assessment of the scatter plot one may notice that the points are not linear. The residual plot may also suggest a definite curvature. This gives a room to fit the nonlinear regression model. Polynomial regression adds extra independent variables that are powers of the original variable, x, x^2, x^3 etc.

The relationship between the response variable Y and the predictor variables X is modeled using a nonlinear function

$$Y = f(X, B) + \varepsilon.$$

where:

f is a nonlinear function of the predictor variables X and parameter β.
β represents the parameters of the nonlinear model.
ε is the random error term.

The general expression for a nonlinear quadratic regression model is given by

$$Y = \beta_0 + \beta_1 X + \beta_2 X_2^2 + \varepsilon.$$

where:

Y is the response variable.
X is the predictor variable.
β_0, β_1 and β_2 are coefficients of the model.
ε is the error term.

The simplest nonlinear regression model is the quadratic equation model where we include the square of the independent variable into the model. This allows the model to capture the curvature in the scatter model.

To perform nonlinear regression in SPSS, we go to: Analyze > Nonlinear > select the dependent variable of interest and move it to the Dependent box and write the expression of interest (e.g. polynomial, exponential etc.) under Model Expression > click Parameters and specify the Name and the value of the parameter in the expression then click Continue then OK. SPSS will display the output for Nonlinear Regression.

5.8 PRACTICE EXERCISE

1. Classify the different generalized linear models (GLMs) techniques for analyzing data.
2. Using relevant SPSS data, perform and interpret the following analysis:
    a. Logistic regression (binary, ordinal, multinomial).
    b. Probit regression.
    c. Poisson regression.
    d. Negative binomial regression.
    e. Loglinear regression.
    f. Gamma.
    g. Inverse Gaussian GLMs.
    h. Nonlinear regression.

CHAPTER 6

Multilevel Modeling

6.1 INTRODUCTION

Multilevel models are particularly appropriate for research designs where data for participants are organized at more than one level (i.e. nested data).

Multilevel modeling or linear mixed models, also known as mixed-effects models or hierarchical linear models, are used to analyze data with hierarchical or nested structures, where observations are grouped into multiple levels. These models account for both fixed effects (i.e. population-level effects) and random effects (i.e. individual/group-specific effects) in the data.

Multilevel modeling, sometimes referred to as hierarchical modeling, is a powerful tool that allows a researcher to account for data collected at multiple levels.

Multilevel modeling insures we get correct coefficients that account for the nesting in the data.

Multilevel models allow us to estimate these different sources of variation.

Hierarchical linear modeling (HLM)/multilevel modeling is a framework for handling nested data structures.

Nested data arise in many disciplines including education, social sciences, medicine and business among others.

Dependence occurs in most of the nested data. Ignoring dependence can cause several problems such as violating the independence assumptions, which may lead to drawing invalid conclusions.

The main benefit of HLMs is that they model dependence nature of data.

DOI: 10.1201/9781003386636-6

6.2 HIERARCHICAL DATA

Data structures for multilevel models are often hierarchical. This is where some variables are clustered or nested within other variables. For example: Children nested within classrooms. This is usually in most cases as a result of using cluster sampling technique. Because of clustering, the assumption of independence is usually violated.

Two-Level Hierarchy: Consider a scenario where children (or case) are organized by classroom. In here, children are said to be nested within classes. The Level 1 variable at the bottom of the hierarchy.

The general expression for a two-level hierarchical model where Level 1 is nested within Level 2 unit is given by, i.e., Level 1 Model (Within-Group Model),

$$Y_{ij} = \beta_{0j} + \beta_{1j}X_{ij} + \varepsilon_{ij}.$$

where:

Y_{ij} is the response variable for the ith individual in the jth group.
X_{ij} is the predictor variable for the ith individual in the jth group.
β_{0j} is the intercept for group j.
β_{1j} is the slope for group j.
ε_{ij} is the Level 1 residual error term.

The Level 2 Model (Between-Group Model) is

$$\beta_{0j} = \gamma_{00} + \mu_{0j}.$$

$$\beta_{1j} = \gamma_{10} + \mu_{1j}.$$

where:

γ_{00} is the overall intercept (fixed effect).
μ_{0j} is the random intercept effect for group j.
γ_{10} is the overall slope (fixed effect).
μ_{1j} is the random slope effect for group j.

The combined model is given by

$$Y_{ij} = (\gamma_{00} + \mu_{0j}) + (\gamma_{10} + \mu_{1j})X_{ij} + \varepsilon_{ij}.$$

Three-Level Hierarchy: There are three levels in the hierarchy: the child (Level 1), the class to which the child belongs (Level 2) and the school within which that class exists (Level 3). Hierarchical data structures need not apply only to between-participant situations. We can also think of data as being nested within people. In this situation, the case, or person, is not at the bottom of the hierarchy (Level 1) but is further up.

The general multilevel model, with *p* predictors at Level 1 and *q* predictors at Level 2 is given by

$$Y_{ij} = \beta_{0j} + \sum_{k=1}^{p} \beta_{kj} X_{kij} + \varepsilon_{ij}.$$

$$\beta_{ij} = \gamma_{k0} + \sum_{l=1}^{q} \gamma_{kl} Z_{lj} + \mu_{kj}.$$

where:

Z_{lj} are the Level 2 predictors.

γ_{kl} are fixed effects.

μ_{kj} are the random effects at Level 2.

X_{kij} is the *k*th predictor variable for the *i*th individual in the *j*th group.

Z_{lj} is the *l*th predictor variable at the group level.

Multilevel data are more complex and don't meet the assumptions of regular linear or generalized linear models.

6.3 INTRACLASS CORRELATION

The intraclass correlation (ICC) is a useful measure for quantifying the level of dependence in the data. The ICC measures the proportion of the total variability in the outcome that is attributable to the classes. ICC is a good measure of whether a contextual variable has an effect on the outcomes. ICC is computed as

$$\text{ICC} = \frac{\text{between group variability}}{\text{between group variability} + \text{within group variability}}$$

$$= \frac{Var(u_{0j})}{Var(u_{0j}) + Var(u_{ij})}.$$

The intraclass correlation coefficient (ICC) is used to calculate the portion of variance in the dependent variable that is explained at each level in subsequent models with the addition of individual and group measures. ICC value is zero.

6.4 VISUALIZING MULTILEVEL DATA

Hierarchical data can be visualized using tree diagram (also referred to as tree map). Tree map consists of a series of nested rectangles with their sizes being proportional to the corresponding data value. We can also use spaghetti plots and heatmaps.

6.5 MULTILEVEL MODELS

Multilevel models, also known as hierarchical linear models or mixed-effects models, are used to analyze data with nested or hierarchical structures, where observations are grouped into multiple levels. These models account for both fixed effects (population-level effects) and random effects (group-specific effects) in the data.

To develop and understand multilevel models, we usually start from the simplest model and build it up to the most complicated model. The process starts from the null model, then to the random-intercept only model (random-intercept model), then to both random-intercept and random-slope models (random-slope model).

Multilevel models are also known as mixed-effects models and random effects models, random-coefficient regression models or covariance components models. The mixed-effects model is because of the mixture of random-intercept and random-slope models.

In order to run a multilevel model, you have to be clear about what is fixed and what is random in your model.

To perform multilevel modeling in SPSS, we go to: Analyze > Mixed Models > Linear > under Subjects, select the ID for Level 3 and Level 2 then click Continue > drag the dependent variable to the Dependent variable box > click Fixed and select build terms then click Continue > click Random and drag Level 3 ID to Combinations then click Next and drag Level 2 ID to Combinations then click Continue > click Estimation then select Maximum Likelihood (ML) and residual method then click Continue > click Statistics and select Parameter estimates for fixed effects, Tests for covariance parameters and Covariances then click Continue then click OK. SPSS Output will display the results of Mixed Model Analysis.

6.5.1 The Null Model

This is the first model in multilevel modeling. It is used to estimate the outcomes without any predictor. It used to determine if the average difference exists in the outcome variable across all the levels.

The null model in a two-level hierarchical structure where individuals (Level 1 units) are nested within groups (Level 2 units) is given by

$$Y_{ij} = \gamma_{00} + \mu_{oj} + \varepsilon_{ij}.$$

where:

γ_{00} is the overall mean intercept.
μ_{oj} is the group-level random effect.
ε_{ij} is the individual-level residual error.

6.5.2 The Random-Intercept Model

The random-intercept model extends the null model to add predictors (i.e. independent variables). Each cluster is assumed to be unique in the random-intercept model.

The random-intercept model in a two-level hierarchical structure where individuals (Level 1 units) are nested within groups (Level 2 units) is given by

$$Y_{ij} = \gamma_{00} + \mu_{oj} + \beta_1 X_{ij} + \varepsilon_{ij}.$$

where:

γ_{00} is the overall mean intercept.
μ_{oj} is the group-level random effect.
β_1 is the fixed slope coefficient for the predictor X_{ij}.
ε_{ij} is the individual-level residual error.

6.5.3 The Random-Slope Model

The random-slope model extends the random-intercept model by adding the random-slope (i.e. random effect for slopes).

The random-slope model in a two-level hierarchical structure where individuals (Level 1 units) are nested within groups (Level 2 units) is given by

$$Y_{ij} = \gamma_{00} + \gamma_{10} X_{ij} + \mu_{oj} + \mu_{1j} X_{ij} + \varepsilon_{ij}.$$

where:

γ_{00} is the overall mean intercept.
γ_{10} is the overall mean slope.
μ_{oj} is the group-level random effect.
μ_{1j} is the group-level random slope effect.
ε_{ij} is the individual-level residual error.

6.5.4 Fitting Two-Level Models

A researcher wants to determine the extent to which vocabulary scores can be used to predict general reading achievement where students are nested within schools.

An example of two-level data: A researcher wants to determine the extent to which vocabulary scores can be used to predict general reading achievement where students are nested within schools. In this case, school is a random effect, and vocabulary scores are fixed.

6.5.5 Fitting Models of Three and More Levels

This involves expanding basic two-level framework by fitting models with additional levels of data structure. The following steps is usually adopted: first, specify the Level 1 equation, then specify the Level 2 equation and finally substitute to have a mixed model.

The general three-level model where individuals (Level 1) are nested within groups (Level 2), which are further nested within larger units (Level 3) is given by

$$Y_{ijk} = \delta_{000} + \delta_{100} X_{ijk} + v_{00k} + v_{10k} X_{ijk} + u_{ojk} + u_{1jk} X_{ijk} + \varepsilon_{ijk}.$$

where:

δ_{000} is the overall mean intercept.
δ_{100} is the overall mean slope.
v_{00k} is the random intercept effect at unit k.
v_{10k} is the random slope effect at unit k.
u_{ojk} is the random intercept effect at group j in unit k.
u_{1jk} is the random slope effect at group j in unit k.
ε_{ijk} is the residual error at individual level.

Example of three-level data: Nurses (Level 1) working within wards (Level 2), with each ward being nested within one of the 25 hospitals (Level 3). Nurses nested within wards and wards nested within the hospital. We would like to determine whether nurses receiving an experimental treatment would differ in their level of stress from those who did not receive the treatment.

The outcome variable measured at Level 1 was stress.

The Level 1 predictors included gender (0 = male, 1 = female), age (in years) and experience (in years).

The Level 2 predictors included experimental condition (0 = control, 1 = experimental), ward type (0 = general care, 1 = special care).

The Level 3 predictors included hospital size (1 = small, 2 = medium, 3 = large).

6.6 RANDOM EFFECTS

In a model with only fixed effects, the residual covariance matrix is diagonal as each observation is assumed independent. In mixed models, there is a dependence structure across observations, so the residual covariance matrix cannot be diagonal.

When fitting a random effects model, effects can be included as either *nested* or *crossed.*

6.6.1 Nested Random Effects

Nested random effects are when each member of one group is contained entirely within a single unit of another group. An example is students in classrooms; you may have repeated measures per student, but each student belongs to a single classroom (assuming no reassignments).

Nested random effects assume that there is some kind of hierarchy in the grouping of the observations, e.g., schools and classes. A class groups a number of students, and a school groups a number of classes. There is a one-to-many relationship between the random effects. For example, a school can contain multiple classes, but a class can only be part of one school.

The general nested random effects model is given by

$$Y_{ij} = \beta_0 + \beta_1 X_{ij} + \upsilon_{oj} + \varepsilon_{ij}.$$

where:

Y_{ij} is the response variable for observation i on group j.
X_{ij} is the predictor variable for observation i in group j.
β_0 and β_1 are the fixed effects for the intercept and slope, respectively.
υ_{oj} is the random effect for group j.
ε_{ij} is the residual error term.

6.6.2 Crossed Random Effects

Crossed random effects are when this nesting is not true. An example would be when we have different seeds and different fields being used for planting crops. Seeds of the same type can be planted in different fields, and each field can have multiple seeds in it.

Crossed random effects appear when two (or more) variables can be used to create distinct groupings. Consider a scenario of factories and products where a factory can produce a range of products, and a product can be manufactured in different factories.

The general nested random effects model is given by

$$Y_{ijk} = \beta_0 + \beta_1 X_{ijk} + u_{0i} + v_{0j} + \varepsilon_{ijk}.$$

where:

Y_{ijk} is the response variable for observation k in the ith level of the first grouping factor and the jth level of the second grouping factor.

X_{ijk} is the predictor variable for observation k in the ith level of the first grouping factor and the jth level of the second grouping factor.

β_0 and β_1 are the fixed effects for the intercept and slope, respectively.

u_{0i} and v_{0j} are ith level of the first grouping factor and the jth level of the second grouping factor, respectively.

ε_{ijk} is the residual error term.

Example 6.1: Using multilevel data (Figure 6.1). (a) Run a random-intercept model, (b) Add in Level 1 predictors (c) and in Level 2 predictors.

We are predicting students posttest scores (Level 1 outcome) from Level 1 and Level 2 predictors. The Level 1 predictors include gender and pretest scores (Figure 6.1).

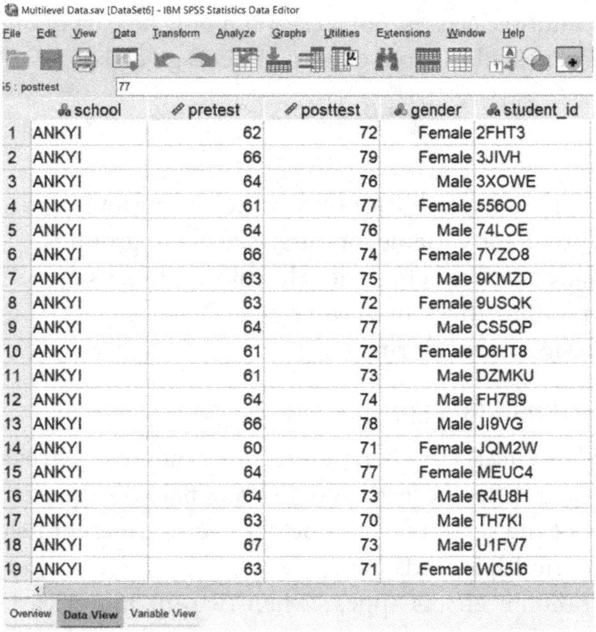

Multilevel Data.sav [DataSet6] - IBM SPSS Statistics Data Editor

File Edit View Data Transform Analyze Graphs Utilities Extensions Window Help

5 : posttest 77

	school	pretest	posttest	gender	student_id
1	ANKYI	62	72	Female	2FHT3
2	ANKYI	66	79	Female	3JIVH
3	ANKYI	64	76	Male	3XOWE
4	ANKYI	61	77	Female	556O0
5	ANKYI	64	76	Male	74LOE
6	ANKYI	66	74	Female	7YZO8
7	ANKYI	63	75	Male	9KMZD
8	ANKYI	63	72	Female	9USQK
9	ANKYI	64	77	Male	CS5QP
10	ANKYI	61	72	Female	D6HT8
11	ANKYI	61	73	Male	DZMKU
12	ANKYI	64	74	Male	FH7B9
13	ANKYI	66	78	Male	JI9VG
14	ANKYI	60	71	Female	JQM2W
15	ANKYI	64	77	Female	MEUC4
16	ANKYI	64	73	Male	R4U8H
17	ANKYI	63	70	Male	TH7KI
18	ANKYI	67	73	Male	U1FV7
19	ANKYI	63	71	Female	WC5I6

Overview Data View Variable View

FIGURE 6.1 Multilevel Data

Model 1: Fitting random intercept model.

This model allows the intercepts to randomly vary between schools because no predictors are included into the model at Level 1.

We go to Analyze > Mixed Models > Linear > move the School Id as the Subjects then click Continue > select the outcome variable (Poststest scores) and move it to the Dependent Variable box > click on Random then click include intercept then move School id to the Combinations box then click Continue > click Estimation and select Maximum Likelihood (ML) then click Continue > click Statistics and select Parameter estimates for fixed effects, Test for covariance effects and Covariances for random effects then click Continue then OK (Figure 6.2).

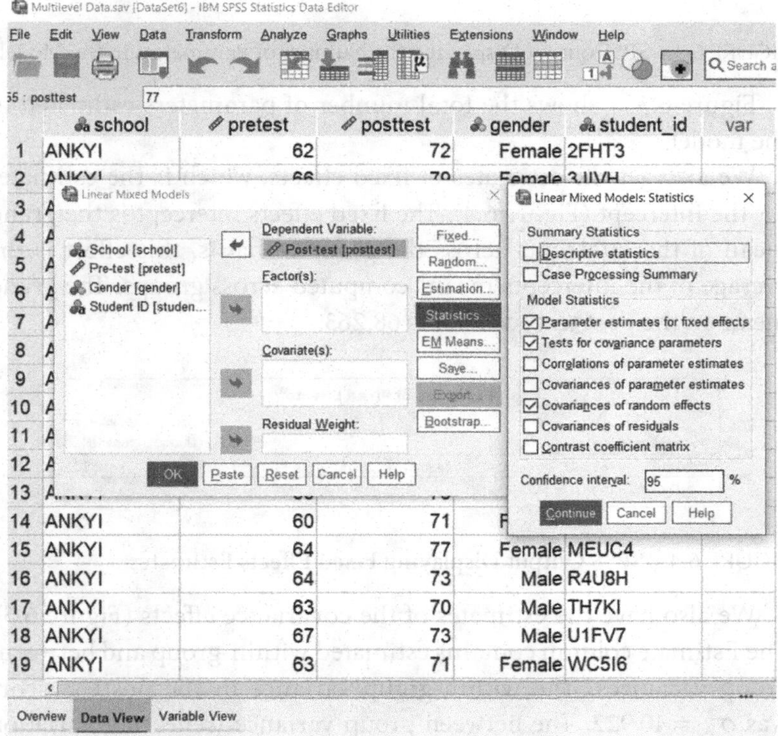

FIGURE 6.2 SPSS Procedure for Fitting Random Intercept Model

SPSS will display Model 1 output.

Model Dimension[a]

		Number of Levels	Covariance Structure	Number of Parameters	Subject Variables
Fixed Effects	Intercept	1		1	
Random Effects	Intercept	1	Variance Components	1	school
Residual				1	
Total		2		3	

a. Dependent Variable: Post-test.

Information Criteria[a]

-2 Log Likelihood	14100.571501
Akaike's Information Criterion (AIC)	14106.571501
Hurvich and Tsai's Criterion (AICC)	14106.582774
Bozdogan's Criterion (CAIC)	14126.567355
Schwarz's Bayesian Criterion (BIC)	14123.567355

The information criteria are displayed in smaller-is-better form.

a. Dependent Variable: Post-test.

FIGURE 6.3 SPSS output Displaying the Number of Parameters in the Model

Figure 6.3 shows the total number of parameters estimated in the model.

We also get the estimates of fixed effects, which is the estimates for the intercept (Figure 6.4). The fixed effects intercept is the grand mean of the intercepts across the groups. That is, this effect is the average of the intercepts/means computed across groups. The grand mean on the post test scores was 68.263.

Estimates of Fixed Effects[a]

Parameter	Estimate	Std. Error	df	t	Sig.	95% Confidence Interval Lower Bound	95% Confidence Interval Upper Bound
Intercept	68.263	2.434	23.000	28.051	<.001	63.229	73.298

a. Dependent Variable: Post-test.

FIGURE 6.4 SPSS Output Displaying Fixed Effects Estimates

We also have the estimates of the covariance effects (Figure 6.5). The Estimate column contains estimated within-group and between-group variances. The within-group variance in the posttest score was $\sigma_W^2 = 40.922$. The between-group variance (reflecting variation in intercepts, which are simply the group means on the dependent variable) was $\sigma_B^2 = 135.705$. These estimates can be tested for statistical significance to determine if there is significant variation to be explained at Levels 1 and 2. The Wald Z tests show that both variance components were statistically significant ($p < 0.001$).

Covariance Parameters

Estimates of Covariance Parameters[a]

Parameter		Estimate	Std. Error	Wald Z	Sig.	95% Confidence Interval Lower Bound	95% Confidence Interval Upper Bound
Residual		40.922	1.260	32.481	<.001	38.526	43.467
Intercept [subject = school]	Variance	135.705	40.167	3.379	<.001	75.972	242.403

a. Dependent Variable: Post-test.

FIGURE 6.5 SPSS Output Displaying the Covariance Effects

To obtain the intraclass correlation coefficient (ICC) allows further evaluation of the level of nonindependence. ICC is the expected correlation between any two randomly chosen individuals in the same group and is computed as the proportion of variation in the Level 1 outcome explained by the grouping structure). ICC values greater than 0.5 are often considered an indicator of nontrivial amount of nonindependence. For the current model ICC was 0.7683.

$$\mathrm{ICC} = \frac{\sigma_B^2}{\sigma_W^2 + \sigma_B^2} = \frac{135.705}{135.705 + 40.922} = 0.7683.$$

Model 2: Random intercept model + addition of Level 1 predictors.

This implies that the model includes predictors for Level 1 that have been group-mean centered, the intercepts/means for groups are adjusted for within-cluster (i.e. schools) association between the Level 1 predictors and outcomes. The intercepts are not adjusted for between-cluster differences on the predictor variables.

We go to Analyze > Mixed Models > Linear then select School id as the Subjects then click Continue then select Post-test score as Dependent variable then click continue > select Pre-test and gender as Covariate(s) > click Fixed and select Pretest and gendewc (mean for Gender) and then click Add then click Continue then OK (Figure 6.6).

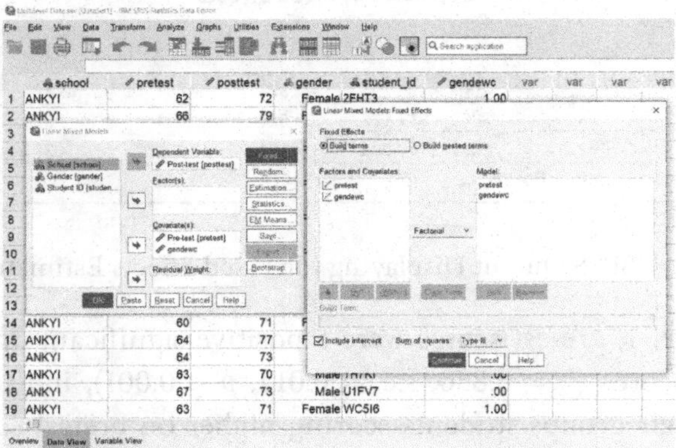

FIGURE 6.6 SPSS Procedure for Fitting Random Intercept Model

SPSS Output will display the results for Model 2 (Figure 6.7).

Mixed Model Analysis

Model Dimension[a]		Number of Levels	Covariance Structure	Number of Parameters	Subject Variables
Fixed Effects	Intercept	1		1	
	pretest	1		1	
	gendewc	1		1	
Random Effects	Intercept	1	Variance Components	1	school
Residual				1	
Total		4		5	

a. Dependent Variable: Post-test.

Information Criteria[a]

-2 Log Likelihood	12124.913479
Akaike's Information Criterion (AIC)	12134.913479
Hurvich and Tsai's Criterion (AICC)	12134.941688
Bozdogan's Criterion (CAIC)	12168.239903
Schwarz's Bayesian Criterion (BIC)	12163.239903

The information criteria are displayed in smaller-is-better form.

a. Dependent Variable: Post-test.

FIGURE 6.7 SPSS output Displaying the Parameter Estimates

From Figure 6.7, we have the total number of parameters estimated in the model to be 5.

We also have the SPSS Outputs for the estimates for the fixed effects (Figure 6.8).

Estimates of Fixed Effects[a]

Parameter	Estimate	Std. Error	df	t	Sig.	95% Confidence Interval Lower Bound	Upper Bound
Intercept	20.652	.923	155.255	22.377	<.001	18.829	22.476
pretest	.850	.014	964.578	60.528	<.001	.823	.878
gendewc	-.022	.178	2108.909	-.123	.902	-.371	.327

a. Dependent Variable: Post-test.

FIGURE 6.8 SPSS Output Displaying the Fixed Effects Estimates

At Level 1, pretest scores were a positive significant predictor of posttest scores ($b = 0.850$, s.e. $= 0.014$, $p < 0.001$), indicating that within their groups students scoring higher on pretest scores were predicted to have higher posttest scores.

We also have gendewc (original gender was coded 0 = male, 1 = female) was a nonsignificant negative predictor (b = −0.022, s.e. = 0.178, $p = 0.902$), indicating that within their groups females tended to score less on post test scores than males.

The regression model will be given by

$$y_{ij} = \gamma_{00} + \mu_{oj} + \gamma_{10} gendecw\gamma_{ij} + \gamma_{20} pretest\gamma_{ij} + e_{ij}.$$

$$y_{ij} = 20.652 + \mu_{oj} - 0.022 gendecw\gamma_{ij} + 0.850 pretest\gamma_{ij} + e_{ij}.$$

We also have the covariance parameters table (Figure 6.9).

Covariance Parameters

Estimates of Covariance Parameters[a]

Parameter		Estimate	Std. Error	Wald Z	Sig.	95% Confidence Interval Lower Bound	Upper Bound
Residual		16.625	.512	32.450	<.001	15.651	17.661
Intercept [subject = school]	Variance	5.006	1.666	3.005	.003	2.607	9.610

a. Dependent Variable: Post-test.

FIGURE 6.9 SPSS Output Displaying Covariance Estimates

The within-group variance in posttest scores was $\sigma_W^2 = 16.625$. The between-group variance (reflecting the variations in intercepts) was $\sigma_B^2 = 5.006$.

Based on the results for the Wald Z tests, both variance components were statistically significant ($p < 0.01$).

We now fit Model 3: Random intercept and Level 1 predictors plus Level 2 predictors.

We have Level 1 equation given by

$$y_{ij} = \beta_{0j} + \beta_{1j} gendecw_{ij} + \beta_{2j} \text{pretest}_{ij} + e_{ij}.$$

Level 2 equation is given by

$$\beta_{0j} = \gamma_{00} + \gamma_{01} gende\_mean_j + \gamma_{02} \text{pretest\_mean}_{ij} + \mu_{0j}.$$

$$\beta_{1j} = \gamma_{10}.$$

$$\beta_{2j} = \gamma_{20}.$$

We go to Analyze > Mixed Models > Linear then select School id as the Subjects then click Continue then select Post-test score as Dependent variable then click continue > select gende_mean, pretest_mean, pre-test and gender as Covariate(s) > click Fixed and select gende_mean, pretest_mean, pretest and gendewc (mean for Gender) and then click Add then click Continue then OK (Figure 6.10).

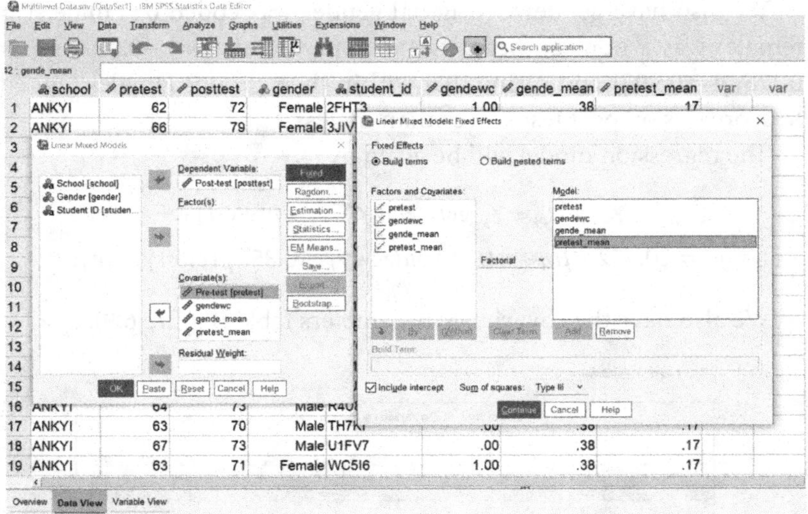

FIGURE 6.10 SPSS Procedure for Fitting Model 3

Mixed Model Analysis

Model Dimension[a]

		Number of Levels	Covariance Structure	Number of Parameters	Subject Variables
Fixed Effects	Intercept	1		1	
	pretest	1		1	
	gendewc	1		1	
	gende_mean	1		1	
	pretest_mean	1		1	
Random Effects	Intercept	1	Variance Components	1	school
Residual				1	
Total		6		7	

a. Dependent Variable: Post-test.

Information Criteria[a]

-2 Log Likelihood	12122.012585
Akaike's Information Criterion (AIC)	12136.012585
Hurvich and Tsai's Criterion (AICC)	12136.065291
Bozdogan's Criterion (CAIC)	12182.669578
Schwarz's Bayesian Criterion (BIC)	12175.669578

The information criteria are displayed in smaller-is-better form.

a. Dependent Variable: Post-test.

FIGURE 6.11 SPSS output Displaying Parameter Estimates

Under the model dimension (Figure 6.11), we have seven parameters being estimated in the model.

We have estimates of fixed effects table (Figure 6.12).

Estimates of Fixed Effects[a]

Parameter	Estimate	Std. Error	df	t	Sig.	95% Confidence Interval Lower Bound	Upper Bound
Intercept	-4.181	412.596	2114.896	-.010	.992	-813.317	804.955
pretest	.850	.014	981.563	60.467	<.001	.822	.877
gendewc	-.026	.178	2108.959	-.148	.882	-.375	.323
gende_mean	1094.020	641.922	2126.670	1.704	.088	-164.839	2352.880
pretest_mean	-2299.218	2767.698	2106.045	-.831	.406	-7726.925	3128.489

a. Dependent Variable: Post-test.

FIGURE 6.12 SPSS Output Displaying Fixed Effects Estimates

Here, at Level 1, pretest (b = 0.850, s.e. = 0.014, $p < 0.001$) is the only positive and significant predictors of post test scores.

At Level 2, none of the variables was a significant predictor of posttests scores ($p > 0.05$).

We also have the covariance parameters table (Figure 6.13).

Covariance Parameters

Estimates of Covariance Parameters[a]

Parameter		Estimate	Std. Error	Wald Z	Sig.	95% Confidence Interval Lower Bound	Upper Bound
Residual		16.600	.512	32.450	<.001	15.627	17.634
Intercept [subject= school]	Variance	5.073	1.685	3.011	.003	2.646	9.726

a. Dependent Variable: Post-test.

FIGURE 6.13 SPSS Output Displaying Fixed Effectes Estimates

The within-group variance in posttest scores was $\sigma_W^2 = 16.60$. The between-group variance (reflecting the variations in intercepts) was $\sigma_B^2 = 5.073$.

Based on the results for the Wald Z tests, both variance components were statistically significant ($p < 0.01$).

The ICC after accounting for Level 1 predictors was 0.234.

$$\text{ICC} = \frac{\sigma_B^2}{\sigma_W^2 + \sigma_B^2} = \frac{5.073}{16.60 + 5.073} = 0.234.$$

This is lower than the random intercept model.

The ICC and test of the variance component for the first model (i.e. random intercept model) as well as for the remaining models suggests a substantial clustering in the data. The findings support the use of HLM and opposed to the use of OLS regression to analyze the data. Even after controlling for the predictors at Levels 1 and 2, there still remained a significant variation left to be explained (based on the tests of the variance components).

6.7 PRACTICE EXERCISE

1. Classify the different multilevel modeling techniques for analyzing data.
2. Using relevant SPSS data, perform and interpret the following analysis:
   a. Random intercept model.
   b. Two-level regression.
   c. Three-level regression.

CHAPTER 7

Longitudinal/Panel Data Analysis

7.1 INTRODUCTION

Longitudinal data are a type of data that are collected over time from the same subjects. This means that the same subjects are observed multiple times, which allows researchers to track how they change over time.

Longitudinal data are often used in psychology, education and medicine to study a variety of phenomena, such as the development of cognitive abilities, the effects of educational interventions and the progression of diseases.

There are two main types of longitudinal data: *repeated measures data* and *panel data. Repeated measures data* are data in which the same variables are measured on the same subjects multiple times. *Panel data* are data in which different variables are measured on the same subjects multiple times.

Longitudinal studies are repeated measurements through time, whereas cross-sectional studies are a single outcome per individual. Observations from an individual tend to be correlated, and the correlation must be taken into account for valid inference.

Longitudinal data analysis involves analyzing data that are collected over multiple time points for the same individuals or entities. This type of data is also known as panel data or repeated measures data. Longitudinal data analysis allows researchers to study changes over time, individual trajectories and the effects of time-varying predictors.

DOI: 10.1201/9781003386636-7

Longitudinal data have repeated observations on individuals, allowing the direct study of change. The measurements for longitudinal data are commensurate i.e. the same variable is measured repeatedly. Longitudinal data require sophisticated statistical techniques because the repeated observations are usually (positively) correlated. The sequential nature of the measures implies that certain types of correlation structures are likely to arise. Correlation must be accounted for to obtain valid inferences.

Longitudinal data are special form of multilevel data where repeated longitudinal data points can be observed for each individual. This means that we can use multilevel techniques to analyze longitudinal data.

7.2 VISUALIZING LONGITUDINAL DATA

There are a number of ways to visualize longitudinal data. Some of the most common techniques include:

Spaghetti plots: Spaghetti plots are a simple way to visualize longitudinal data. A spaghetti plot is a line plot with a separate line for each subject. The lines are connected to show the change in the variable (or relationship – linear or nonlinear) over time.

Growth curves: Growth curves are a more sophisticated way to visualize longitudinal data. A growth curve is a line plot with a smooth curve fitted to the data. The curve can be used to show the average change in the variable over time, as well as the variation in the data around the average.

Boxplots: Boxplots can be used to visualize the distribution of the variable over time. A boxplot shows the median, quartiles and outliers for the variable at each time point.

Heatmaps: Heatmaps can be used to visualize the correlation between the variable and time. A heatmap shows the correlation between the variable and time as a color scale.

To generate Spaghetti Plot in SPSS, go to: Analyze > General Linear Model > Repeated Measures > specify the number of level items you would like to specify then click Define and move the selected variable to Within-Subjects Variables box > add ID to Between-Subjects factors box> click profile Plots then click create plots and drag factor 1 to Horizontal Axis box and ID to Separate Plots box then click Add then click Continue then click OK. SPSS will generate a Spaghetti plot.

Example 7.1: Using spaghetti data (Figure 7.1), generate a spaghetti plot for five levels.

Spagetti Data.sav [DataSet2] - IBM SPSS Statistics Data Editor

	ID	time_1	time_2	time_3	time_4	time_5	time_6
1	1.00	10.13	9.46	5.98	45.26	8.77	53.55
2	2.00	4.97	5.03	6.28	31.33	5.36	7.17
3	3.00	8.10	6.67	5.67	62.94	5.89	47.20
4	4.00	10.57	8.12	17.00	95.14	5.04	53.20
5	5.00	7.92	5.87	16.23	73.82	6.58	31.43
6	6.00	9.18	6.83	5.31	64.78	9.50	25.51
7	7.00	5.56	9.98	7.43	60.00	5.09	10.59
8	8.00	4.51	5.54	5.51	40.34	8.76	6.07
9	9.00	6.23	5.66	7.85	17.88	5.89	37.48
10							

FIGURE 7.1 Spaghetti Data

In SPSS, we go to: Analyze > General Linear Model > Repeated Measures > specify the number of level items you would like to specify (5) then click Define and move the five selected variables (time_1, time_2, time_4, time_6, time_7) to Within-Subjects Variables box > add ID to Between-Subjects factors box > click profile Plots then click create plots and drag factor 1 to Horizontal Axis box and ID to Separate Plots box then click Add then click Continue then click OK (Figure 7.2).

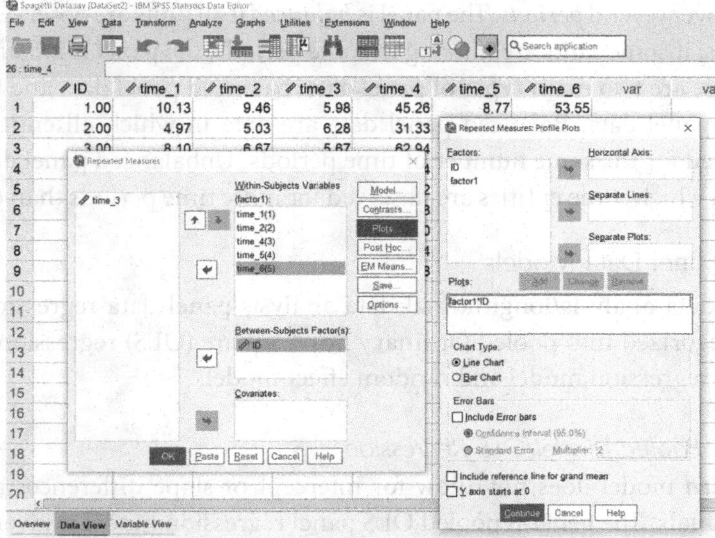

FIGURE 7.2 SPSS procedure for Generating Spaghetti Plot

SPSS will generate a Spaghetti plot (Figure 7.3).

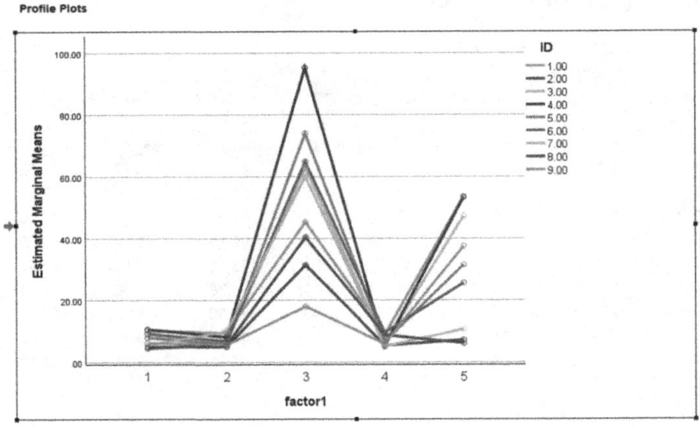

FIGURE 7.3 SPSS Output Displaying Spaghetti Plot

7.3 PANEL DATA

Panel data are a type of data that contain observations about the same entities (individuals, households, firms, countries, etc.) over time. This means that each entity is observed multiple times, which allows researchers to track how the entities change over time and how they interact with each other.

Panel data gather information about several individuals (cross-sectional *units*) over several *periods*. The panel is *balanced* if all units are observed in all periods; if some units are missing in some periods, the panel is *unbalanced*.

There are two main types of panel data: balanced panel data and unbalanced panel data. Balanced panel data are data in which all entities are observed for the same number of time periods. Unbalanced panel data are data in which some entities are observed for more time periods than others.

7.3.1 Panel Data Models

Panel data analysis/longitudinal data analysis/panel data regression can be categorized into pooled Ordinary Least Square (OLS) regression, fixed effects regression model and random effect model.

7.3.1.1 Pooled Model/OLS Regression

A *pooled* model does not allow for intercept or slope differences among individuals. The general pooled OLS panel regression model is given by

$$Y_{it} = \beta_0 + \beta_1 X_{1it} + \beta_2 X_{2it} + \ldots + \beta_k X_{kit} + \varepsilon_{it}.$$

where:

Y_{it} is the dependent variable for entity *i* at time *t*.
β_0 is the constant/intercept.
$X_{1it}, X_{2it}, \ldots, X_{kit}$ are the dependent variables for entity *i* at time *t*.
$\beta_1, \beta_2, \ldots, \beta_k$ are the coefficients in the independent variables.
ε_{it} is the error term.

7.3.1.2 The Fixed Effects Model

The fixed effects model (FEM) takes into account individual differences, translated into different intercepts of the regression line for different individuals.

The general fixed effects model for panel data is given by

$$Y_{it} = \beta_0 + \beta_1 X_{1it} + \beta_2 X_{2it} + \ldots + \beta_k X_{kit} + \alpha_i + \varepsilon_{it}.$$

where:

Y_{it} is the dependent variable for entity *i* at time *t*.
β_0 is the constant/intercept.
$X_{1it}, X_{2it}, \ldots, X_{kit}$ are the dependent variables for entity *i* at time *t*.
$\beta_1, \beta_2, \ldots, \beta_k$ are the coefficients in the independent variables.
α_i is the entry-specific fixed effect.
ε_{it} is the error term.

7.3.1.3 The Random Effects Model

The *random effects* model (REM) elaborates on the fixed effects model by recognizing that, since the individuals in the panel are randomly selected, their characteristics, measured by the intercept, which should also be random.

The general random effects model for panel data is given by

$$Y_{it} = \beta_0 + \beta_1 X_{1it} + \beta_2 X_{2it} + \ldots + \beta_k X_{kit} + \alpha_i + \varepsilon_{it}.$$

where:

Y_{it} is the dependent variable for entity *i* at time *t*.
β_0 is the constant/intercept.
$X_{1it}, X_{2it}, \ldots, X_{kit}$ are the dependent variables for entity *i* at time *t*.
$\beta_1, \beta_2, \ldots, \beta_k$ are the coefficients in the independent variables.
α_i is the entry-specific random effect.
ε_{it} is the error term.

To perform panel regression analysis in SPSS, we go to:

OLS Model

Analyze > Regression > Linear > select the dependent variable of interest and move it to the Dependent variable then select the independent variable of interest and move it to the independent variable box then click OK. SPSS Output will generate the OLS results.

Go to Data > Split File > click Compare groups then select year and move it to Groups Based on box then click OK. Analyze > Regression > Linear > select the dependent variable of interest and move it to the Dependent variable then select the independent variable of interest and move it to the independent variable box then click OK. SPSS Output will generate the regression results per year.

The fixed effects regression analysis can be performed using least square dummy variable approach.

The first step is to ensure that the data are in long format using SPSS path: Data > Restructure.

To perform fixed effects regression analysis can be performed using least square dummy variable approach:

We first create a dummy variable for each observation. We go to: Transform > Create Dummy Variables > move the ID variable to Create Dummy Variables for: box and type dum in the Root Names box then click OK. SPSS will create dummy variable for each case > we the perform regression by including the number of dummy variables less 1.

We go to Analyze > Regression > Linear > move the dependent variable into the Dependent box. Move the dummy variables (less 1) into the Independent(s) box then click the Next button and add the remaining independent variables into Block 2 of 2 box > click Statistics and select R square change then click Continue then OK. SPSS will give the regression output.

Example 7.2: Using the panel data (Figure 7.4), run a fixed effects regression.

	no	ID	subject	state	yr	sales	resale	price
1	1	1	1	1	1992	16.919	16.360	21.500
2	8	1	1	2	1992	51.102	16.725	21.200
3	15	1	1	3	1992	63.729	22.525	39.895
4	22	1	1	4	1992	24.629	10.310	18.890
5	29	1	1	5	1992	7.854	12.360	19.840
6	36	1	1	6	1992	76.034	7.750	12.640
7	43	1	1	7	1992	31.038	13.425	18.575
8	50	1	1	8	1992	245.815	10.055	17.885
9	57	1	1	9	1992	540.561	15.075	26.935
10	64	1	1	10	1992	66.692	7.825	11.799
11	71	1	1	11	1992	24.072	26.975	31.505
12	78	1	1	12	1992	48.911	21.725	43.330
13	2	2	2	1	1993	39.384	19.875	28.400
14	9	2	2	2	1993	9.231	28.675	33.400
15	16	2	2	3	1993	15.943	27.100	44.475
16	23	2	2	4	1993	42.593	11.525	19.390
17	30	2	2	5	1993	32.775	14.180	24.495
18	37	2	2	6	1993	4.734	12.545	19.045
19	44	2	2	7	1993	111.313	11.260	16.980
20	51	2	2	8	1993	42.574	17.810	29.299

FIGURE 7.4 Panel Data

We create a dummy variable for each observation. We go to: Transform > Create Dummy Variables > move the ID variable to Create Dummy Variables for: box and type dum in the Root Names box then click OK (Figure 7.5).

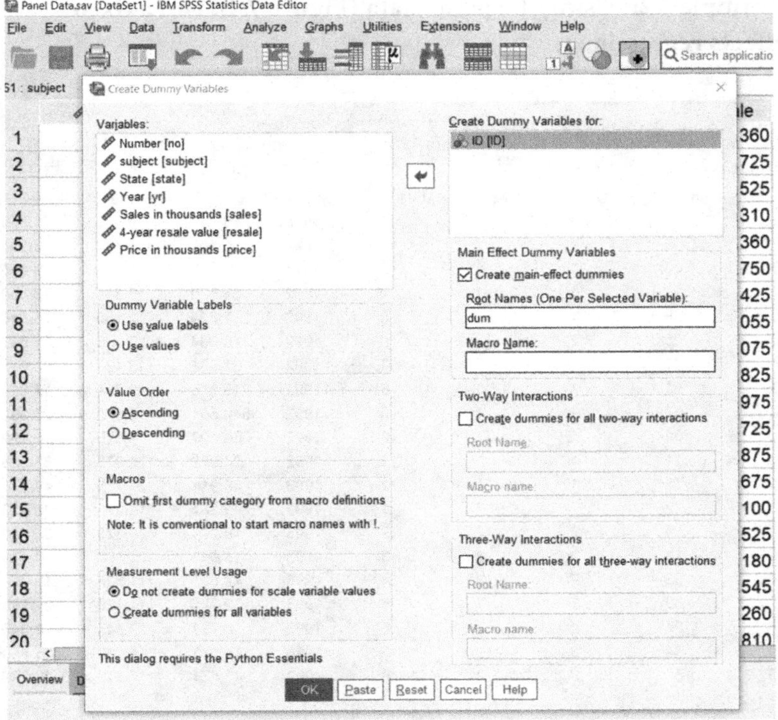

FIGURE 7.5 SPSS Procedure for Creating Dummy Variables in Panel Regression Analysis

SPSS will create dummy variable for each case (Figure 7.6).

*Panel Data.sav [DataSet1] - IBM SPSS Statistics Data Editor

	no	ID	subject	state	yr	sales	resale	price	dum_1	dum_2	dum_3
1	1	1	1	1	1992	16.919	16.360	21.500	1.00	.00	.00
2	8	1	1	2	1992	51.102	16.725	21.200	1.00	.00	.00
3	15	1	1	3	1992	63.729	22.525	39.895	1.00	.00	.00
4	22	1	1	4	1992	24.629	10.310	18.890	1.00	.00	.00
5	29	1	1	5	1992	7.854	12.360	19.840	1.00	.00	.00
6	36	1	1	6	1992	76.034	7.750	12.640	1.00	.00	.00
7	43	1	1	7	1992	31.038	13.425	18.575	1.00	.00	.00
8	50	1	1	8	1992	245.815	10.055	17.885	1.00	.00	.00
9	57	1	1	9	1992	540.561	15.075	26.935	1.00	.00	.00
10	64	1	1	10	1992	66.692	7.825	11.799	1.00	.00	.00
11	71	1	1	11	1992	24.072	26.975	31.505	1.00	.00	.00
12	78	1	1	12	1992	48.911	21.725	43.330	1.00	.00	.00
13	2	2	2	1	1993	39.384	19.875	28.400	.00	1.00	.00
14	9	2	2	2	1993	9.231	28.675	33.400	.00	1.00	.00
15	16	2	2	3	1993	15.943	27.100	44.475	.00	1.00	.00
16	23	2	2	4	1993	42.593	11.525	19.390	.00	1.00	.00
17	30	2	2	5	1993	32.775	14.180	24.495	.00	1.00	.00
18	37	2	2	6	1993	4.734	12.545	19.045	.00	1.00	.00
19	44	2	2	7	1993	111.313	11.260	16.980	.00	1.00	.00
20	51	2	2	8	1993	42.574	17.810	29.299	.00	1.00	.00

FIGURE 7.6 SPSS Data View Window Displaying the Dummy Variables

While performing regression, we include the number of dummy variables less 1.

We go to Analyze > Regression > Linear > move the dependent variable (Sales) into the Dependent box. Move the six dummy variables into the Independent(s) box then click the Next button and add the remaining independent variables (sales, resale) into Block 2 of 2 box > click Statistics and select R square change then click Continue then OK (Figure 7.7).

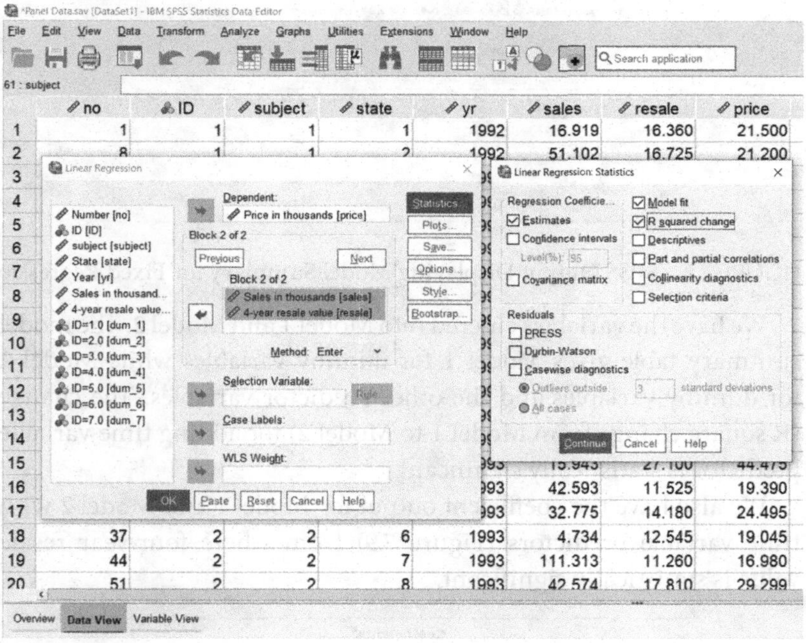

FIGURE 7.7 SPSS procedure for Fixed Regression Analysis

Variables Entered/Removed[a]

Model	Variables Entered	Variables Removed	Method
1	ID=7.0, ID=2.0, ID=6.0, ID=4.0, ID=5.0, ID=3.0[b]	.	Enter
2	Sales in thousands, 4-year resale value[b]	.	Enter

a. Dependent Variable: Price in thousands

b. All requested variables entered.

Model Summary

Model	R	R Square	Adjusted R Square	Std. Error of the Estimate	Change Statistics				
					R Square Change	F Change	df1	df2	Sig. F Change
1	.254[a]	.065	-.008	10.257461	.065	.885	6	77	.510
2	.923[b]	.851	.835	4.146573	.787	198.093	2	75	<.001

a. Predictors: (Constant), ID=7.0, ID=2.0, ID=6.0, ID=4.0, ID=5.0, ID=3.0

b. Predictors: (Constant), ID=7.0, ID=2.0, ID=6.0, ID=4.0, ID=5.0, ID=3.0, Sales in thousands, 4-year resale value

ANOVA[a]

Model		Sum of Squares	df	Mean Square	F	Sig.
1	Regression	558.784	6	93.131	.885	.510[b]
	Residual	8101.594	77	105.216		
	Total	8660.378	83			
2	Regression	7370.823	8	921.353	53.586	<.001[c]
	Residual	1289.555	75	17.194		
	Total	8660.378	83			

a. Dependent Variable: Price in thousands

b. Predictors: (Constant), ID=7.0, ID=2.0, ID=6.0, ID=4.0, ID=5.0, ID=3.0

FIGURE 7.8 SPSS Output Displaying Model Summary for Fixed Regression

We have the variables entered into Model 1 and Model 2. The model summary table gives Model 1 for dummy variables while Model 2 for dummy variables and the other predictor variables. The change (R square change from Model 1 to Model 2) for adding time variable predictor is statistically significant.

We also have the coefficient output for Model 1 and Model 2 with time variable predictors (Figure 7.9) from where four-year resale value is statistically significant.

Coefficients[a]

Model		Unstandardized Coefficients B	Unstandardized Coefficients Std. Error	Standardized Coefficients Beta	t	Sig.
1	(Constant)	23.666	2.961		7.992	<.001
	ID=2.0	.671	4.392	.021	.153	.879
	ID=3.0	1.892	4.035	.069	.469	.640
	ID=4.0	4.776	4.188	.165	1.141	.258
	ID=5.0	-4.405	4.188	-.152	-1.052	.296
	ID=6.0	-.821	4.188	-.028	-.196	.845
	ID=7.0	.525	4.188	.018	.125	.901
2	(Constant)	4.491	1.752		2.563	.012
	ID=2.0	-.592	1.796	-.019	-.329	.743
	ID=3.0	-1.253	1.657	-.046	-.756	.452
	ID=4.0	-.559	1.717	-.019	-.326	.746
	ID=5.0	-1.671	1.735	-.058	-.963	.338
	ID=6.0	-.510	1.696	-.018	-.301	.765
	ID=7.0	.159	1.700	.005	.093	.926
	Sales in thousands	.002	.006	.016	.344	.732
	4-year resale value	1.257	.065	.923	19.253	<.001

a. Dependent Variable: Price in thousands

FIGURE 7.9 SPSS Output Displayng the Coefficients from Fixed Regression

7.4 REPEATED MEASURES DATA

Repeated measures data come from experiments where you take observations repeatedly over time. Under a repeated measures experiment, experimental units are observed at multiple points in time. So instead of looking at an observation at one point in time, we will look at data from more than one point in time. With this type of data, we are looking at only a single response variable but measured over time.

The term repeated measures refers to experimental designs (or observational studies) in which each experimental unit (or subject) is measured at several points in time.

One approach to analyzing these data is the ANOVA model.

7.4.1 Repeated Measures ANOVA

The Repeated measures ANOVA is a statistical technique that is used to analyze data on the same subjects measured multiple times. Repeated measures ANOVA can be used to test for the effects of a treatment or intervention on a group of subjects over time.

The general form of the repeated measures ANOVA is given by

$$Y_{ij} = \mu + \alpha_i + \beta_j + (\alpha \times \beta)_{ij} + \varepsilon_{ij}.$$

where:

Y_{ij} is the observed value for the dependent variable for subject i at time or condition j.

μ is the overall mean of the dependent variable.

α_i is the effect of the ith subject.

β_j is the effect of the jth condition.

$(\alpha \times \beta)_{ij}$ is the interaction effect between subjects and conditions/time points.

ε_{ij} is the residual error term.

To run ANOVA with repeated measures using SPSS, we go to: Click Analyze > General Linear Model > Repeated measures > type Time under Within-Subject Factor Name box and indicate the number of time levels under Number of levels box then click Add then click Define > move the first variable to Time Level 1, second variable to Time Level 2, third variable to Time Level 3 and so on depending under the Within-Subjects Variables > click Plots and move Time to Horizontal Axis box then click Add and Continue > click Options and select Descriptive statistics and Estimates of effect size then click

Continue then OK. SPSS will display the output for repeated measures ANOVA.

Example 7.3: Perform one-way repeated measures ANOVA for the following data (Figure 7.10).

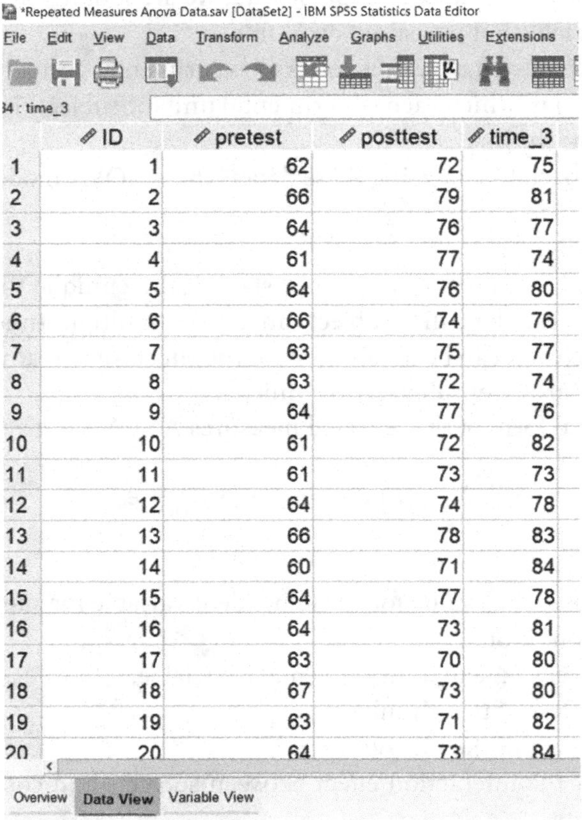

*Repeated Measures Anova Data.sav [DataSet2] - IBM SPSS Statistics Data Editor

File Edit View Data Transform Analyze Graphs Utilities Extensions

34 : time_3

	ID	pretest	posttest	time_3
1	1	62	72	75
2	2	66	79	81
3	3	64	76	77
4	4	61	77	74
5	5	64	76	80
6	6	66	74	76
7	7	63	75	77
8	8	63	72	74
9	9	64	77	76
10	10	61	72	82
11	11	61	73	73
12	12	64	74	78
13	13	66	78	83
14	14	60	71	84
15	15	64	77	78
16	16	64	73	81
17	17	63	70	80
18	18	67	73	80
19	19	63	71	82
20	20	64	73	84

Overview Data View Variable View

FIGURE 7.10 Repeated Measures Data

We go to: Click Analyze > General Linear Model > Repeated measures > type Time under Within-Subject Factor Name box and 3 under Number of levels box then click Add then click Define > move pretest to Time Level 1, Post-test to Time Level 2 and time after 3 months to Time Level 3 under the Within-Subjects Variables > click Plots and move Time to Horizontal Axis box then click Add and Continue > click Options and select Descriptive statistics and Estimates of effect size then click Continue then OK (Figure 7.11).

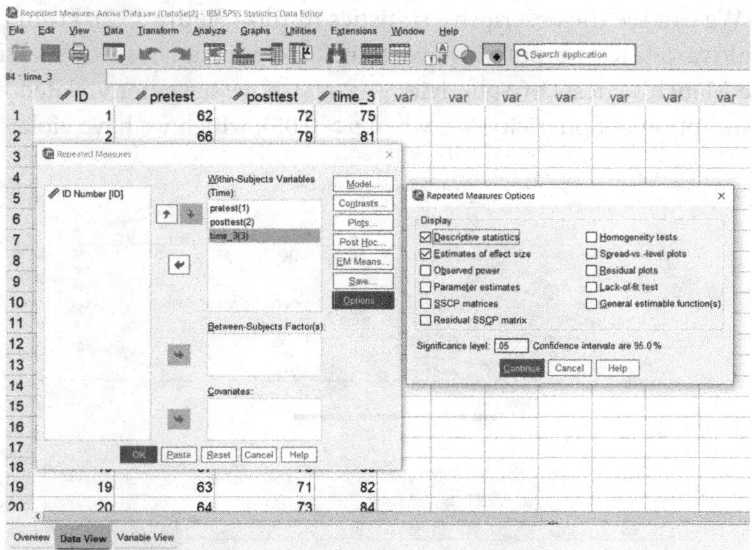

FIGURE 7.11 SPSS Procedure for Repeated Measures ANOVA

SPSS will display the output for repeated measures ANOVA (Figure 7.12).

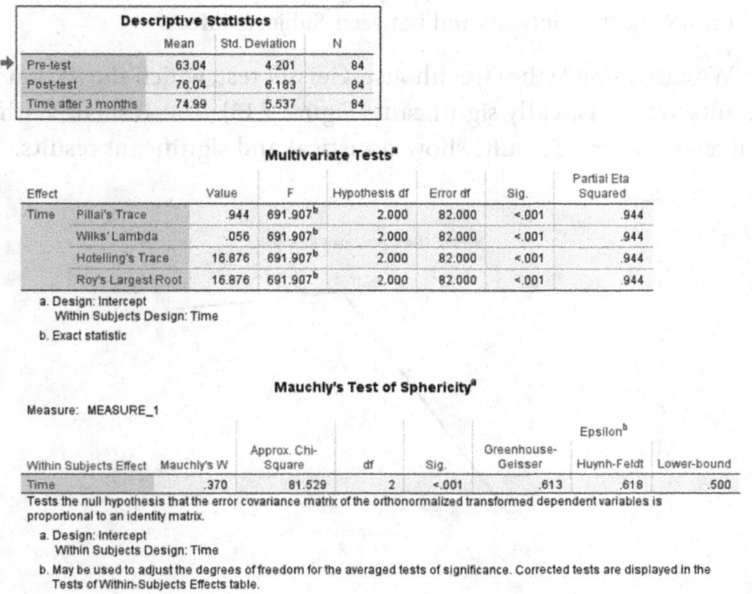

Descriptive Statistics

	Mean	Std. Deviation	N
Pre-test	63.04	4.201	84
Post-test	76.04	6.183	84
Time after 3 months	74.99	5.537	84

Multivariate Tests[a]

Effect		Value	F	Hypothesis df	Error df	Sig.	Partial Eta Squared
Time	Pillai's Trace	.944	691.907[b]	2.000	82.000	<.001	.944
	Wilks' Lambda	.056	691.907[b]	2.000	82.000	<.001	.944
	Hotelling's Trace	16.876	691.907[b]	2.000	82.000	<.001	.944
	Roy's Largest Root	16.876	691.907[b]	2.000	82.000	<.001	.944

a. Design: Intercept
Within Subjects Design: Time

b. Exact statistic

Mauchly's Test of Sphericity[a]

Measure: MEASURE_1

Within Subjects Effect	Mauchly's W	Approx. Chi-Square	df	Sig.	Epsilon[b] Greenhouse-Geisser	Huynh-Feldt	Lower-bound
Time	.370	81.529	2	<.001	.613	.618	.500

Tests the null hypothesis that the error covariance matrix of the orthonormalized transformed dependent variables is proportional to an identity matrix.

a. Design: Intercept
Within Subjects Design: Time

b. May be used to adjust the degrees of freedom for the averaged tests of significance. Corrected tests are displayed in the Tests of Within-Subjects Effects table.

FIGURE 7.12 SPSS Output Displaying Discriptive Statistics and Multivariate

We can see the descriptive statistics results (Figure 7.12). From the multivariate test output, we have statistical significant results. We have the Mauchly's tests of Sphericity to assess if we have not violated the assumption of Sphericity (i.e. when $p > 0.05$), which we have violated.

Tests of Within-Subjects Effects

Measure: MEASURE_1

Source		Type III Sum of Squares	df	Mean Square	F	Sig.	Partial Eta Squared
Time	Sphericity Assumed	8762.794	2	4381.397	151.317	<.001	.646
	Greenhouse-Geisser	8762.794	1.227	7141.677	151.317	<.001	.646
	Huynh-Feldt	8762.794	1.236	7089.916	151.317	<.001	.646
	Lower-bound	8762.794	1.000	8762.794	151.317	<.001	.646
Error(Time)	Sphericity Assumed	4806.540	166	28.955			
	Greenhouse-Geisser	4806.540	101.840	47.197			
	Huynh-Feldt	4806.540	102.584	46.855			
	Lower-bound	4806.540	83.000	57.910			

Tests of Within-Subjects Contrasts

Measure: MEASURE_1

Source	Time	Type III Sum of Squares	df	Mean Square	F	Sig.	Partial Eta Squared
Time	Linear	6000.095	1	6000.095	195.151	<.001	.702
	Quadratic	2762.698	1	2762.698	101.703	<.001	.551
Error(Time)	Linear	2551.905	83	30.746			
	Quadratic	2254.635	83	27.164			

Tests of Between-Subjects Effects

Measure: MEASURE_1
Transformed Variable: Average

Source	Type III Sum of Squares	df	Mean Square	F	Sig.	Partial Eta Squared
Intercept	1283001.433	1	1283001.433	44814.237	<.001	.998
Error	2376.234	83	28.629			

FIGURE 7.13 SPSS Output Displaying Tests for Within-Subject Effects, Within-Subjects Contrasts and Between-Subjects Effects

We can now use the Greenhouse-Geisser test, which shows that the results are statistically significant (Figure 7.13). The Tests of Within-Subjects contrasts results show statistical and significant results.

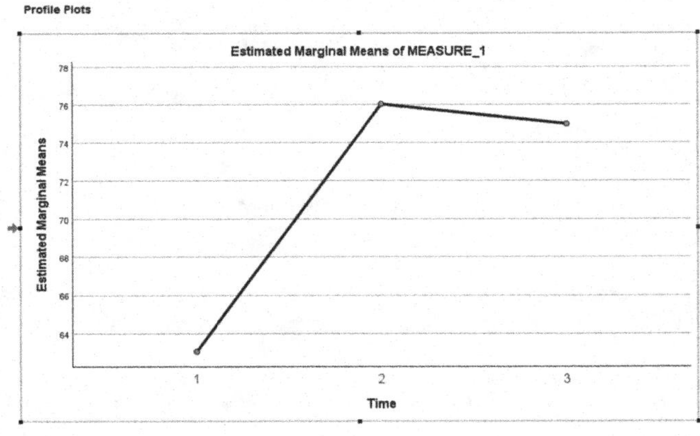

FIGURE 7.14 Profile Plots for Estimated Marginal Means

We move to the results of the estimated marginal means (Figure 7.14), which shows that pretest scores were lower before it goes up to posttest and starts dropping 3 months after posttest.

7.5 GENERALIZED ESTIMATING EQUATIONS

Generalized estimating equation (GEE) can take into account the correlation of within-subject data (longitudinal studies) and other studies in which data are clustered within subgroups. Generalized estimating equations can be thought of as an extension of generalized linear models (GLM) to longitudinal data. Instead of attempting to model the within-subject covariance structure, GEE models the average response. The goal is to make inferences about the population when accounting for the within-subject correlation. This technique extends the generalized linear model to include analysis of repeated measurements or other correlated observations. For example, it allows for within-subject covariates or autocorrelated response data, such as autoregressive data. The possible types of response variables are the same as the generalized linear models where the response variable involve link functions. Like the generalized linear models procedure, the main dialog box has tabs at the top. The first tab at top is "Repeated". In repeated tab, specify the subject variables and within-subject variables and the structure of working correlation matrix.

7.5.1 Correlation Matrices

The goal of specifying a working correlation structure is to estimate B more efficiently. Incorrect specification can affect efficiency of the parameter estimates.

Autoregressive correlation structure: We consider data that are correlated within clusters over time. In this case, within-subject correlations are set as an exponential function of this lag period that are determined by researcher.

Exchangeable: In this case, within-subject observations are equally correlated. No logical ordering for observations within a cluster-usually appropriate for data that are clustered within a subject but are not time-series data. It is also known as compound symmetry correlation matrix. This is the most often used correlation structure.

Unstructured: In this case, we have free estimation on the within-subject correlation. The correlation matrix estimates all possible correlations between within-subject responses and includes them in the estimation of the variances.

Independence: In this case, there is no correlation within subjects.

GEE works with many types of data. It has the repeated measures aspect. GEE is the most flexible model among the regression models. We can use both continuous and categorical outcome variable.

The general form of GEE model is given by

$$g(\mu_{ij}) = X_{ij}\beta + Z_{ij}\gamma.$$

where:

$g(.)$ is the link function.

μ_{ij} is the expected value of the response variable for observation i in cluster j.

X_{ij} is the matrix of covariates (fixed effects) for observation i in cluster j.

β is the vector of fixed-effect coefficients.

Z_{ij} is the matrix of cluster-specific covariates (random effects) for observation i in cluster j.

γ is the vector of random-effect coefficients.

In SPSS, we go to: Analyze > Generalized Linear Models > Generalized Estimating Equations > under Repeated Tab, select ID under Subject variable and numeric predictor variable under Within-subject variables > under working Correlation Matrix select Exchangeable > under Type of Model Tab select Linear > under Response tab select outcome numeric variable under the Dependent Variable box > under Predictors Tab, select categorical predictor variable as Factors and numeric predictor variable as Covariates. Click Options and select Descending > under Model Tab, select categorical predictor, numeric predictor and Interaction of categorical and numeric predictor and select Interaction under Type > under Statistics Tab also tick include exponential parameter estimates then click OK, and SPSS will give the GEE outputs.

Look at the Parameter Estimates Table and interpret the Beta coefficients are the same as the mixed effect model.

7.5.2 Analyzing Missing Data with Weighted GEE

Missing data frequently occur in longitudinal studies, where missing observations can be caused by dropouts or skipped visits. To draw valid inferences when data are missing, you can use different approaches, such as maximum likelihood, multiple imputation, fully Bayesian analysis and inverse probability weighting. The GEE procedure provides a *weighted*

GEE method for analyzing longitudinal data that have missing observations. This approach extends the usual generalized estimating equations approach. When none of the data are missing, the weighted GEE method is identical to the usual GEE method. The standard GEE method is valid if the data are missing completely at random (MCAR), but it can lead to biased results if the data are missing at random (MAR). The GEE procedure implements the inverse probability-weighted method to account for dropouts under the MAR assumption.

Example 7.4: Perform GEE for the following data (Figure 7.15).

	ID	gender	age	income
1	1	Female	39	42.00
2	2	Male	43	39.00
3	3	Male	49	46.00
4	4	Male	35	17.00
5	5	Male	32	42.00
6	6	Female	21	29.00
7	7	Female	79	9.00
8	8	Male	64	69.00
9	9	Male	62	20.00
10	10	Male	61	105.00
11	11	Female	48	94.00
12	12	Female	35	38.00
13	13	Female	25	35.00
14	14	Female	61	106.00
15	15	Male	57	156.00
16	16	Male	52	59.00
17	17	Female	19	29.00
18	18	Male	65	76.00
19	19	Female	22	20.00
20	20	Female	31	37.00

FIGURE 7.15 GEE Data

In SPSS, we go to: Analyze > Generalized Linear Models > Generalized Estimating Equations > under Repeated Tab, select ID under Subject variable and age under Within-subject variables > under working Correlation Matrix, select Exchangeable > under Type of Model Tab select Linear > under Response tab, select Income (numeric variable) under the Dependent Variable box > under Predictors Tab, select Sex as Factors and Age as Covariates. Click Options and select Descending > under Model Tab select Sex, Age and Interaction of Sex and Age and select Interaction under Type > under Statistics Tab also tick include exponential parameter estimates then click OK (Figure 7.16).

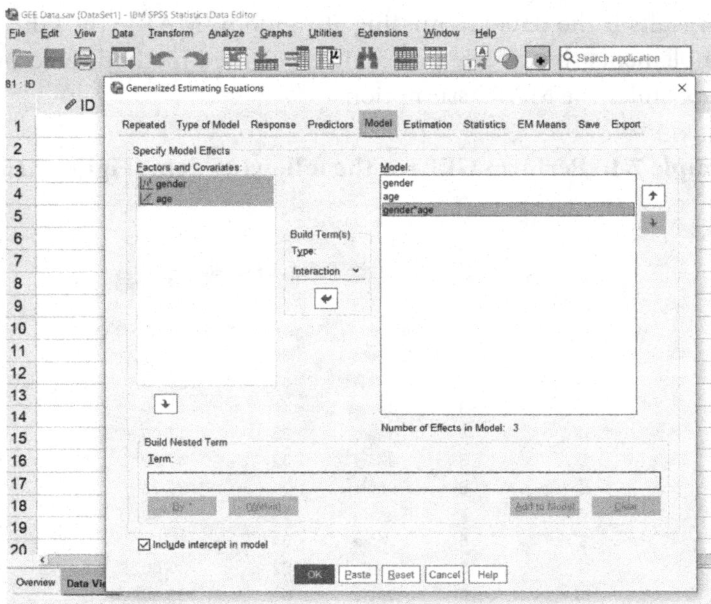

FIGURE 7.16 SPSS Procedure for GEE

SPSS will give the GEE outputs for model information, tests of model effects and parameter estimates (Figure 7.17).

Tests of Model Effects

	Type III		
Source	Wald Chi-Square	df	Sig.
(Intercept)	7.004	1	.008
Gender	2.011	1	.156
Age in years	8.976	1	.003
Gender * Age in years	2.536	1	.111

Dependent Variable: Household income in thousands
Model: (Intercept), Gender, Age in years, Gender * Age in years

Parameter Estimates

			95% Wald Confidence Interval		Hypothesis Test				95% Wald Confidence Interval for Exp(B)	
Parameter	B	Std. Error	Lower	Upper	Wald Chi-Square	df	Sig.	Exp(B)	Lower	Upper
(Intercept)	9.909	11.8915	-13.398	33.216	.694	1	.405	20117.569	1.519E-6	2.665E+14
[Gender=1]	22.877	16.1324	-8.742	54.495	2.011	1	.156	8612965250.4	.000	4.646E+23
[Gender=0]	0[a]	.	.	.	.	.	.	1	.	.
Age in years	1.034	.3198	.407	1.661	10.460	1	.001	2.813	1.503	5.264
[Gender=1] * Age in years	-.718	.4508	-1.601	.166	2.536	1	.111	.488	.202	1.180
[Gender=0] * Age in years	0[a]	.	.	.	.	.	.	1	.	.
(Scale)	2089.725									

Dependent Variable: Household income in thousands
Model: (Intercept), Gender, Age in years, Gender * Age in years
a. Set to zero because this parameter is redundant.

FIGURE 7.17 SPSS Output Displaying the Parameter Estimates from GEE

From the Parameter Estimates Table, age in years is statistically significant.

7.6 PRACTICE EXERCISE

1. Classify the longitudinal/panel techniques for analysing data.
2. Using relevant SPSS data, perform and interpret the following analysis:
    a. Pooled regression.
    b. Fixed effects regression.
    c. Random effect regression.
    d. ANOVA with repeated measures.
    e. Generalized Estimating Equation.

CHAPTER 8

Structural Equation Modeling

8.1 INTRODUCTION

Structural equation modeling (SEM) is a multivariate statistical analysis technique that is used to analyze structural relationships. SEM is a statistical method that examines the relationships among numerous variables in a simultaneous way. It is an extremely broad and flexible framework for data analysis, perhaps better thought of as a family of related methods rather than as a single technique. This technique is the combination of factor analysis and multiple regression analysis, and it is used to analyze the structural relationship between measured variables and latent constructs. This method is preferred by the researcher because it estimates the multiple and interrelated dependence in a single analysis. In this analysis, two types of variables are used endogenous variables and exogenous variables. Endogenous variables are equivalent to dependent variables and are equal to the independent variable.

SEM in a single analysis can assess the assumed causation among a set of dependent and independent constructs i.e. validation of the structural model and the loadings of observed items (measurements) on their expected latent variables (constructs) i.e. validation of the measurement model. The combined analysis of the measurement and the structural model enables the measurement errors of the observed variables to be

DOI: 10.1201/9781003386636-8

analyzed as an integral part of the model, and factor analysis combined in one operation with the hypotheses testing. A construct is measured through a set of items in the questionnaire. Ordinary least squares (OLS) procedures could not entertain latent constructs, hence the need to employ SEM for analysis. Using SEM, we can model the relationship among these constructs together with their respective items in the model and analyze them simultaneously where at least two measurement models are involved i.e. one for independent construct and the other for dependent construct. The structural model is the theorized link between measurement model for independent construct and measurement model for dependent construct.

SEMs differ from other modeling approaches as they test the direct and indirect effects on pre-assumed causal relationships.

A structural equation model (SEM) is a combination of confirmatory factor analysis and path analysis. Structural equation modeling includes two sets of models: the measurement model and the structural model. A measurement model measures the latent variables or composite variables (i.e. expressed as a factor model), while the structural model tests all the hypothetical dependencies based on path analysis.

SEM uses covariance matrix as its input, so you are essentially looking at correlations between variables to determine how one influences the other, but it will not determine causation. SEM is a causal modeling approach. The minimum sample size for SEM is between 100 and 500 samples depending on the number of constructs.

Example 8.1: We can consider a questionnaire comprising of constructs and items corresponding to each of the construct. A construct is measured using a number of different items (usually between four and six items). Construct is considered as latent variables.

Consider a questionnaire in Table 8.1 with four variables for job security (q6), career development (q7), job flexibility (q8) and employee retention (q9).

Job security and career development are constructs. Latent/unobserved variables are measured using the statements/items for constructs.

TABLE 8.1 Sample Questionnaire

Job Security	**1**	**2**	**3**	**4**	**5**
This organization ensure that workers are earning a fair job compensation.					
The perception of job security positively affects retention.					
This organization offers long-term contracts.					
This organization has long-term funding.					
In this organization, your employment can be terminated at any time.					
In this organization, there is a clear process for termination.					
Career Development	**1**	**2**	**3**	**4**	**5**
This organization provides training opportunities to enhance my skills.					
This organization provides opportunities for promotions.					
This organization encourages competitive process for career growth.					
There are opportunities for alternative learning in this organization.					
This organization provides a clear development plan for employee advancement.					
This organization has managers that encourage professional development.					
The rotation management process encourages continuous improvement.					

Note: 1 = strongly disagree, 2 = disagree, 3 = neutral, 4 = agree and 5= strongly agree.

Once the data are collected, it is entered into SPSS (Figure 8.1) and then perform analysis using SEM. The first step is the measurement model.

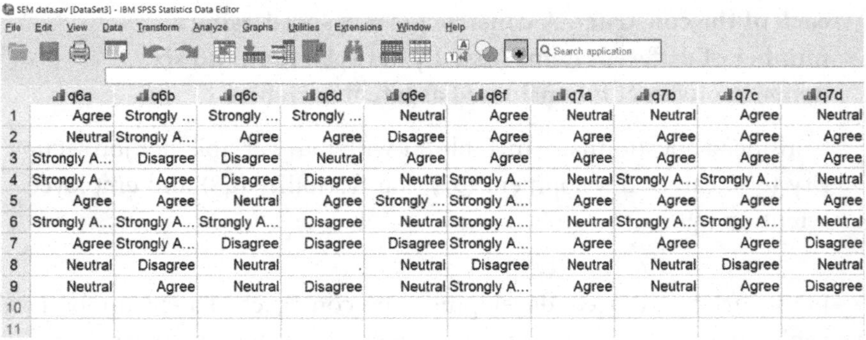

	q6a	q6b	q6c	q6d	q6e	q6f	q7a	q7b	q7c	q7d
1	Agree	Strongly ...	Strongly ...	Strongly ...	Neutral	Agree	Neutral	Neutral	Agree	Neutral
2	Neutral	Strongly A...	Agree	Agree	Disagree	Agree	Agree	Agree	Agree	Agree
3	Strongly A...	Disagree	Disagree	Neutral	Agree	Agree	Agree	Agree	Agree	Agree
4	Strongly A...	Agree	Disagree	Disagree	Neutral	Strongly A...	Neutral	Strongly A...	Strongly A...	Neutral
5	Agree	Agree	Neutral	Agree	Strongly ...	Strongly A...	Agree	Agree	Agree	Agree
6	Strongly A...	Strongly A...	Strongly A...	Disagree	Neutral	Strongly A...	Neutral	Strongly A...	Strongly A...	Neutral
7	Agree	Strongly A...	Disagree	Disagree	Disagree	Strongly A...	Agree	Agree	Agree	Disagree
8	Neutral	Disagree	Neutral	.	Neutral	Disagree	Neutral	Neutral	Disagree	Neutral
9	Neutral	Agree	Neutral	Disagree	Neutral	Strongly A...	Agree	Neutral	Agree	Disagree
10										
11										

FIGURE 8.1 SEM Data

KEY TERMINOLOGIES IN SEM

Constructs: They measure theoretical concepts that are abstract, complex and cannot be directly observed by means of (multiple) items. Constructs are represented in path models as circles or ovals and are also referred to as latent variables e.g. job security, career development from the questionnaire.

Indicators: These are directly measured observations (raw data), also referred to as either items or manifest variables, which are represented in path models as rectangle. They are also available in the data (e.g. the responses to the questions in the questionnaire). They are used in measurement models to measure the latent variables.

Factor loadings: These are the bivariate correlations between a construct and the indicators. They determine an item's absolute contribution to its assigned construct. Loadings are of primary interest in the evaluation of reflective measurement models but are also interpreted when formative measures are involved. The closer the factor loading to 1 the better the representation.

Latent variables: These are elements of a structural model that are used to represent theoretical concepts in statistical models. A latent variable that only explains other latent variables (only outgoing relationships in the structural model) is called exogenous, while latent variables with at least one incoming relationship in the structural model are called endogenous.)

Critical t value: These are the cutoff or the criterion on which the statistical significance of a coefficient is determined. If the computed t value is greater than the t critical value (default is 1.96), the null hypothesis of no effect is rejected.

8.2 CONFIRMATORY FACTOR ANALYSIS (MEASUREMENT MODEL)

The measurement model in SEM is where one assesses the validity of the indicators for each construct. Once you have established the validity of the measurement model, then you can proceed to the structural model.

Confirmatory factor analysis (CFA) is the method for measuring latent variables. Confirmatory factor analysis has an objective to estimate the latent variables. It extracts the latent construct from other variables and shares the most variance with related variables. Confirmatory factor analysis estimates latent variables based on the correlated variations of the dataset.

CFA is a statistical technique that analyzes how well your indicators measure your unobserved constructs, and if your unobserved constructs are uniquely different form one another. In CFA, we treat all the unobserved variables as exogenous or independent variables.

In addition to CFA, there is another type of factor analysis: exploratory factor analysis (EFA). The CFA is applied when the indicator for each latent variable is specified according to the related theories or prior knowledge, whereas EFA is applied to find the underlying latent variables. In practice, EFA is often performed to select the useful underlying latent constructs for CFA when there is little prior knowledge about the latent construct.

CFA can also be done using SPSS-AMOS as part of the structural equation modeling approach. The reliability and validity of the model are assessed using four different values i.e. convergent validity, internal consistency, composite reliability and discriminant validity.

Average variance extracted (AVE) is the measure for understanding convergent validity i.e. construct's ability to share items or statements used to depict it. If the value of AVE for all the variables is more than 0.5 then the model has convergent validity.

Composite reliability (CR) is the method for assessing the contribution or significance of an item by examining the factors loading. If the value of CR is also more than 0.7 for all the constructs, then composite reliability is derived for the model.

Internal consistency is the reliability method for depicting the factor's linkage with other factors. Cronbach's alpha is the method to measure internal consistency. If the value is more than 0.7 for all the variables, then there is the presence of internal consistency in the model.

Discriminant validity is the method for identifying the construct distinction from one another. The value of construct correlation is compared with the square root of AVE. If the correlation value is less than the square root, then the model has discriminant validity.

EFA is useful in data reduction of large number of indicators and can be quite helpful in seeing if indicators are measuring more than one construct. EFA is the first step in determining if an indicator is measuring a construct. In an EFA, one is not denoting which indicators are measuring the construct.

Factor loadings in a CFA estimate the direct effects of unobservable constructs on their indicators. If you have standardized factor loading that is greater than 0.70 or explains at least half of the variance in the indicator then your indicator is providing value in explaining the unobserved construct.

8.3 PATH ANALYSIS (STRUCTURAL MODEL)

The structural model in SEM is concerned with the influence and statistical significance between the constructs.

Path analysis is aimed to find the causal relationship among variables by creating a path diagram. Path analysis was developed to quantify the relationships among multiple variables. Path analysis can explain the causal relationships among variables. A common function of path analysis is mediation, which assumes that a variable can influence an outcome directly and indirectly through another variable. Path analysis may be referred to as regression. Another common function of path analysis is moderation analysis (Figure 8.2).

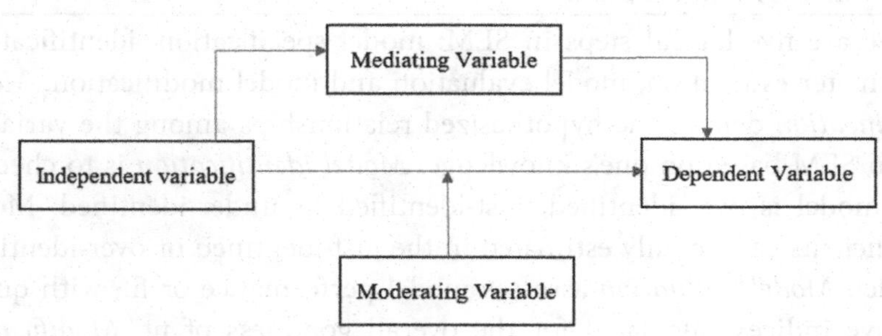

FIGURE 8.2 Mediation and Moderation Analysis

8.3.1 Mediation Analysis

The aim of mediation analysis is to determine if the effect of the independent variable (intervention) on the outcome can be mediated by a change in the mediating variable. Mediation analysis tests a hypothetical causal chain where one variable X affects a second variable M and, in turn, that variable affects a third variable Y. Mediation tests whether the effects of X (the independent variable) on Y (the dependent variable) operate through a third variable, M (the mediator)

8.3.2 Moderation Analysis

Moderation analysis also allows you to test for the influence of a third variable, Z, on the relationship between variables X and Y instead of testing a causal link between these other variables, moderation tests for when or under what conditions an effect occurs. Moderators can strength, weaken or reverse the nature of a relationship. Moderation tests whether a variable (Z) affects the direction and/or strength of the relation between an independent variable (X) and a dependent variable (Y). In other words, moderation tests for interactions that affect when relationships between variables occur.

8.3.2.1 Latent and Observable Variables

SEM applies a confirmatory factor analysis to estimate latent constructs. The latent variable or construct is not in the data set, as it is a derived common factor of other variables and could indicate a model's cause or effect. Latent variables are not directly measurable.

8.4 PERFORMING SEM

There are five logical steps in SEM: model specification, identification, parameter estimation, model evaluation and model modification. *Model specification* defines the hypothesized relationships among the variables in an SEM based on one's knowledge. *Model identification* is to check if the model is over-identified, just-identified or under-identified. Model coefficients can be only estimated in the just-identified or over-identified model. *Model evaluation* assesses model performance or fit, with quantitative indices calculated for the overall goodness of fit. *Modification* adjusts the model to improve model fit i.e. the post hoc model modification. Validation is the process to improve the reliability and stability of the model. Popular programs for SEM applications are often equipped with intuitive manuals, such as AMOS, Mplus, LISREI, Lavaan (R-package), piecewiseSEM (R-package) and Matlab.

Once you have your data (Figure 8.1), the first step in SEM analysis is Measurement Model. Before testing hypothesis, you have to test the measures to test whether they are reliable and valid using measurement model/a covariance-based model. AMOS is a covariance-based software. Measurement model is the confirmatory factor analysis (CFA). CFA is a statistical technique for analyzing how well your indicators measure your unobserved constructs and if your unobserved constructs are uniquely different from one another.

The first step in SEM using AMOS is to draw a **Measurement Model for Confirmatory Factor Analysis**. All the constructs from your questionnaire are drawn and then covaried with each other to perform CFA. CFA is performed by adding all the construct in the questionnaire whether independent, dependent, moderating, endogenous or exogenous. Then the model is assessed for model fit, and the reliability and validity of the constructs is assessed.

Once CFA has been done (i.e. you have established reliability, validity of all the constructs etc.), the next step is the **Structural Model** where you are testing the statistical significance of the hypothesized paths.

8.5 MODEL EVALUATION INDICES

SEM evaluation is based on the fit indices for the test of a single path coefficient (i.e. *p* value and standard error) and the overall model fit (i.e. χ^2, RMSEA).

Chi-square test (χ^2): χ^2 tests the hypothesis that there is a discrepancy between model-implied covariance matrix and the original covariance matrix. Therefore, the nonsignificant discrepancy is preferred i.e. the chi-square value should not be statistically significant if there is a good model fit. A relative chi-square test with a value under 3 is considered acceptable fit.

Root mean square error of approximation (RMSEA) and standardized root mean square residual (SRMR): RMSEA is a "badness of fit" index where 0 indicates the perfect fit, and higher values indicate the lack of fit. The acceptable RMSEA should be less than 0.06 to have a good model fit. SRMR of 0.05 and below will be considered and good fit.

Comparative fit index (CFI): CFI represents the amount of variance that has been accounted for in a covariance matrix. It ranges from 0.0 to 1.0. A higher CFI value indicates a better model fit. In practice, the CFI should be close to 0.95 or higher.

Incremental fit index (IFI): IFI should be 0.90 or higher to have an acceptable fit. IFI is relatively independent to sample size and is one that is frequently reported.

Tucker-Lewis index (TLI): TLI is a non-normed fit index (NNFI) that partly overcomes the disadvantages of normed fit index (NFI) and also proposes a fit index independent of sample size. A TLI of >0.90 is considered acceptable.

Akaike information criterion (AIC) and Bayesian information criterion (BIC): AIC and BIC are two relative measures from the perspectives of model selection rather than the null hypothesis test. AIC offers a relative estimation of the information lost when the given model is used to generate data. BIC is an estimation of how parsimonious a model is among several candidate models.

8.6 INTRODUCTION TO AMOS

AMOS is an abbreviation for analysis of moment structures. Moment refers to means, variance and covariance.

AMOS is a program tied to SPSS that uses a graphical interface for input.

Using AMOS, one can specify, estimate, assess and present the model in a casual path diagram to show the hypothesized relationships among constructs of interest.

The empirical model can be tested against the hypothesized model for goodness of fit.

SEM uses diagrams to denote relationships to be tested. The unobservable constructs are drown using circular or oval shapes. Observed variable indicators are drawn using rectangular boxes. The measurement error/residual error term represents the unexplained variance by the indicator and is drawn using circular with one-way arrow. The direct path effect is represented by a line with a single arrowhead.

We will use SPSS-AMOS to perform SEM.

The first step is to install SPSS-AMOS statistical software.

8.6.1 Steps for Performing Structural Equation Modeling (SEM) Analysis

To Open IBM SPSS Amos.

Once you have installed the IBM SPSS Amos software, click the windows button and search for IBM SPSS Amos folder > click IBM SPSS Amos 29 Graphics > it will open Amos program (Figure 8.3) from where one can draw unobserved constructs and variables. Alternatively, search amos and select IBM SPSS Amos 29 Graphics.

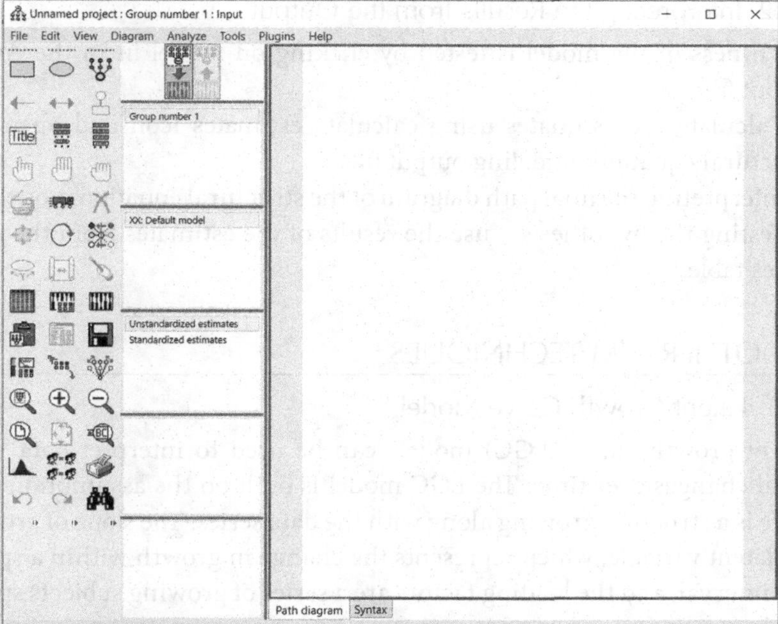

The steps below explains how to perform analysis using SPSS-AMOS:

Step 1: Open IBM SPSS Amos and save the file by selecting File > Save.

Step 2: Import the SPSS data set by selecting "Data Files" from the menu. Select File Name > location of file > file > open > Ok.

Step 3: Draw the path diagram using the draw latent or its indicator icon i.e. for independent and dependent variable.

Step 4: Specify each variable using the imported data set. Drag each variable from variable dialogue box to the drawn observed variable boxes and name the variables.

Step 5: Name all the unobserved variables i.e. residual and measurement error by clicking on Plugins > Name Unobserved Variables.

Step 6: Finally click on the calculate estimates icon to calculate the estimates. A new result file will be created at the location where you saved the Amos file. Open the file.

8.6.2 Interpreting the Results from the Output

The fitness of the model is tested by clicking on model fit in the Amos output file.

Calculate the estimates using calculate estimates icon and open the structural equation modeling output file.

Interpreting the final path diagram of the structural equation modeling.

Testing the hypothesis – use the results of the estimates from the estimates table.

8.7 OTHER SEM TECHNIQUES

8.7.1 Latent Growth Curve Model

Latent growth curve (LGC) models can be used to interpret data with serial changes over time. The LGC model is built on the assumption that there is a structure growing along with the data series. The slope of growth is a latent variable, which represents the change in growth within a specified interval, and the loading factors are a series of growing subjects specified by the user.

8.7.2 Bayesian SEM (BSEM)

Bayesian SEM (BSEM) assumes theoretical support and the prior beliefs are strong. One can use new data to update a prior model so that posterior parameters can be estimated.

8.7.3 Partial Least Square SEM (PLS-SEM)

Partial least square SEM (PLS-SEM) is the preferred method when the study object does not have a well-developed theoretical base, particularly when there is little prior knowledge on causal relationship. The emphasis here is about the explorations rather than confirmations. PLS-SEM requires neither a large sample size nor a specific assumption on the distribution of the data, or even the missing data.

8.7.4 Hierarchical SEM

The hierarchical model, also known as multilevel SEM, analyzes hierarchically clustered data. Hierarchical SEM can specify the direct and indirect causal effect between clusters. It is common for an experiment to fix some variables constantly, resulting in multiple groups or a nested data set. The conventional SEM omits the fact that path coefficients and intercepts will potentially vary between hierarchical levels.

Example 8.2: Using SEM data, determine the relationship between job security (q6), career development (q7), job flexibility (q8) and employee retention (q9); perform SEM by developing measurement model and the structural model.

The measurement model in SEM is where one assesses the validity of the indicators for each construct. After showing the validity of the measurement model, one progresses to the structural model.

The structural model is concerned with the influence and significance between constructs.

Parameters indicate the size and nature of the relationship between two objects in a model.

Search amos and select IBM SPSS Amos 29 Graphics.
Open Amos Software and import data into Amos software.

Performing CFA.

Job security (q6) is hypothesized to positively influence employee retention (q9).

Job security has six indicators while employee retention has nine indicators.

The first step is to design the measurement model.

Click Select data files > click File Name and go to the folder that has SEM data then click OK and the data set will be imported into Amos. To view the data set, click List variables in the data set, and Amos will display the variables in the data (Figure 8.4).

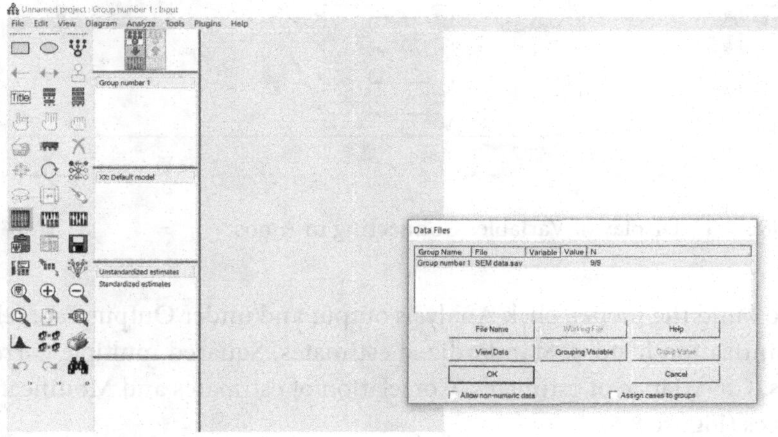

FIGURE 8.4 Display of Variables in Amos

Click Draw a latent variable and click on the empty space. Press the circular shape six times to have six indicators for job security and nine times for employee retention from the SEM data (Figure 8.5).

Click Rotate the indicators of a latent variable icon then bring the mouse to the indicator diagram and click twice to rearrange the indicators.

Click on Preserve symmetry then Move objects icon then click on the indictor diagram to move them to the center of the "white" space.

Click the variable and drag it to the indicator box. Thereafter click Plugins and select Name Unobservable Variables to give the errors the labels. Double click to name the job security (JS) and employee retention (ER). Then, we correlate the two variables using the double arrow (Figure 8.5).

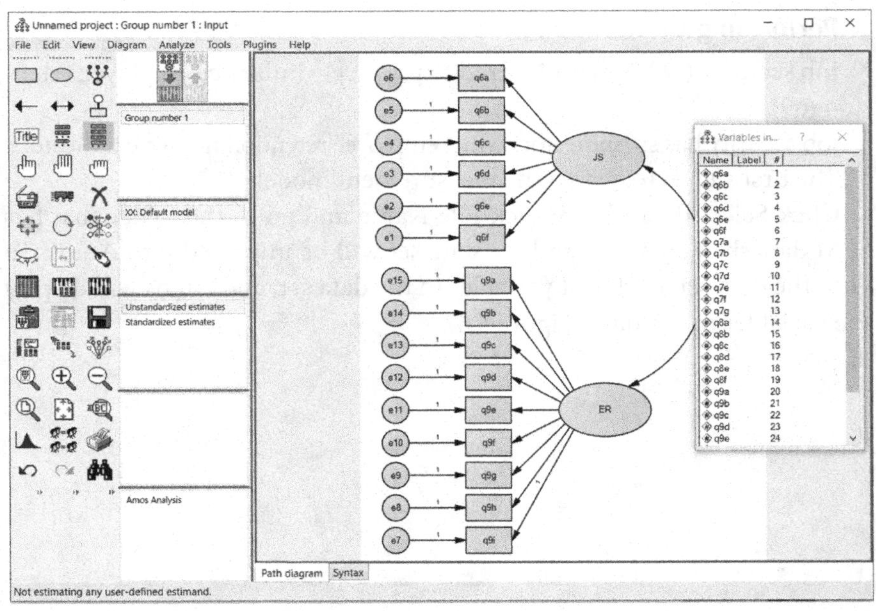

FIGURE 8.5 Display of Variables Connecting in Amos

To build the model, click Analysis output and under Output Tab, select Minimization history, Standardized estimates, Squared multiple correlations, Covariance of estimates, Correlation of estimates and Modification indices (Figure 8.6).

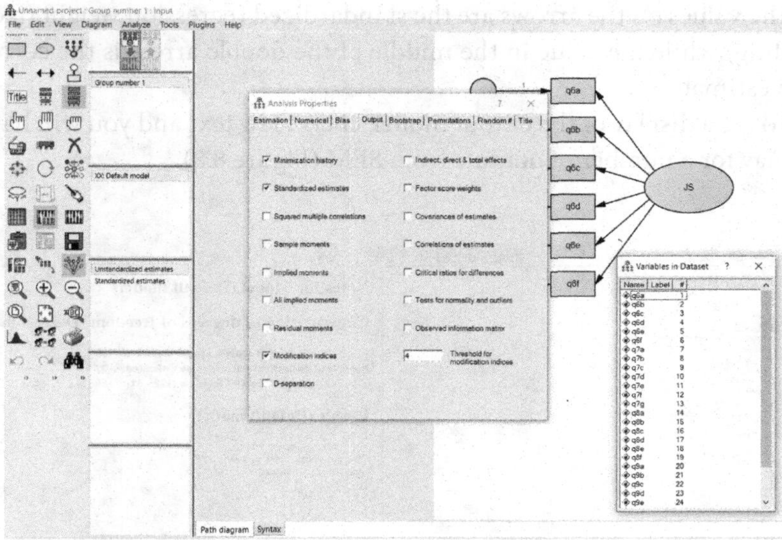

FIGURE 8.6 Amos Procedure for Model building

Then run the model by clicking Calculate the Estimates.

Click File > Save and choose a folder for Amos Analysis. Then Amos will give the CFA results (Figure 8.7).

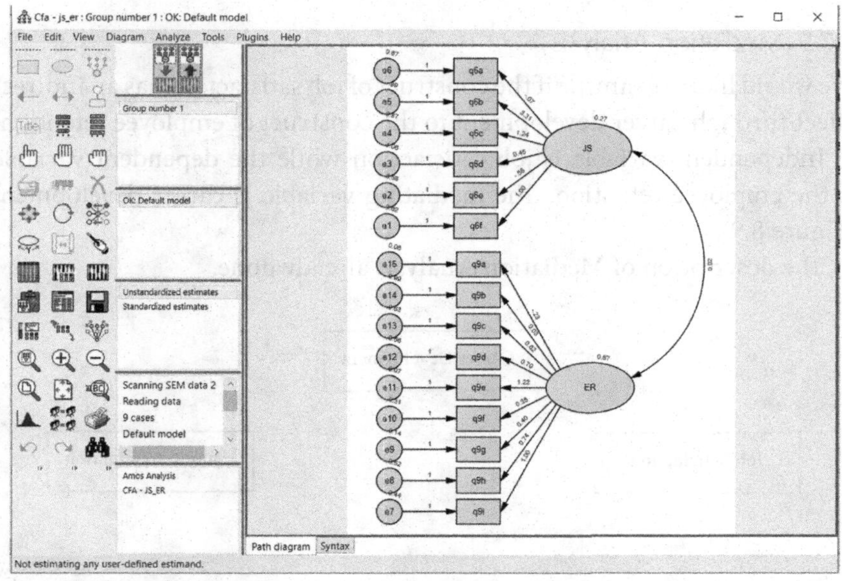

FIGURE 8.7 CFA Results in Amos

The values on the arrows are the standardized regression weight (factor loading) while the value in the middle of the double arrow is the correlation estimate.

To get a display of the output model: click View text and you will have a display for a number of outputs from SEM (Figure 8.8).

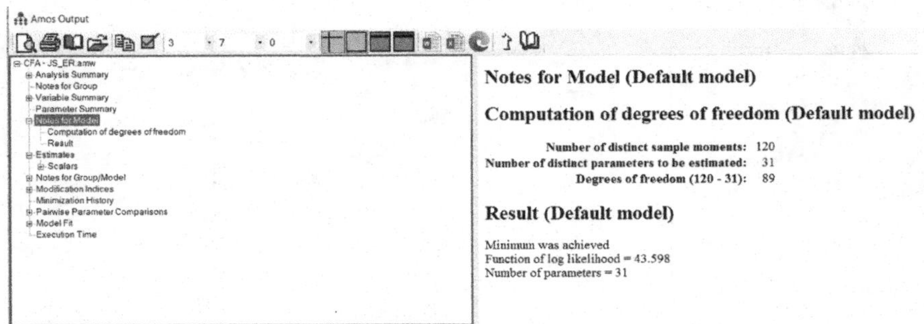

We have the notes for the model that gives the chi-square test, variable summary, parameter summary (with regression weight – Critical Ratio (CR) are the t-values, covariance and correlations), modification indices and model fit summary.

8.7.5 Mediation Analysis

We would like to examine if the construct of job satisfaction has an indirect effect through career development to the construct of employee retention.

Independent variable is job satisfaction while the dependent variable is the employee retention. The mediating variable is career development (Figure 8.9).

The description of Mediation Analysis already done.

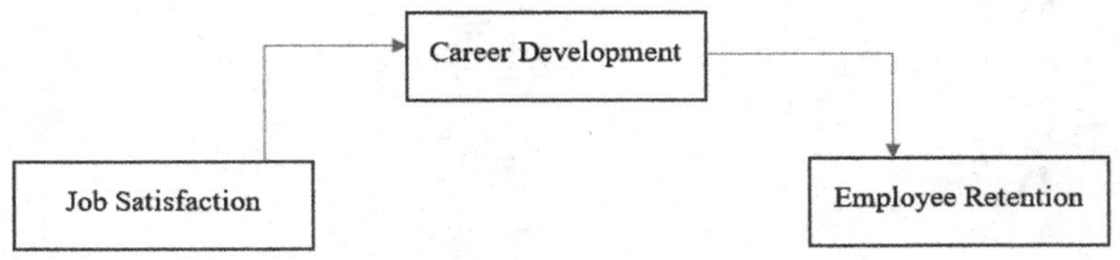

We go to amos software and generate the variables (Figure 8.10).

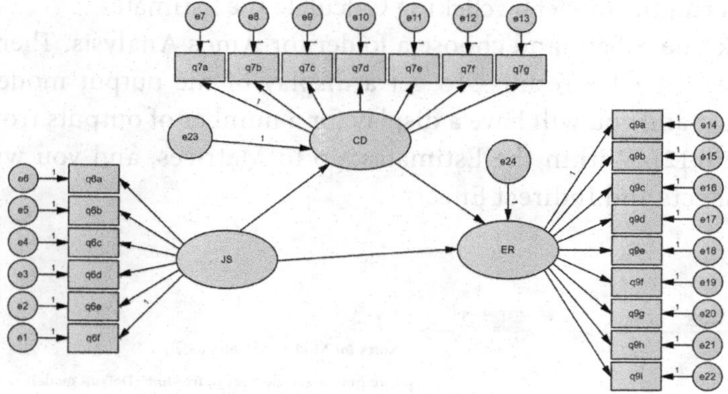

Click Analysis output and under Output Tab, select Minimization history, Standardized estimates, Squared multiple correlations, Covariance of estimates, indirect, direct and total effects, Residual moments, Correlation of estimates and Modification indices. Click Bootstrap tab and select Preform bootstrap and Bias-corrected confidence intervals (Figure 8.11).

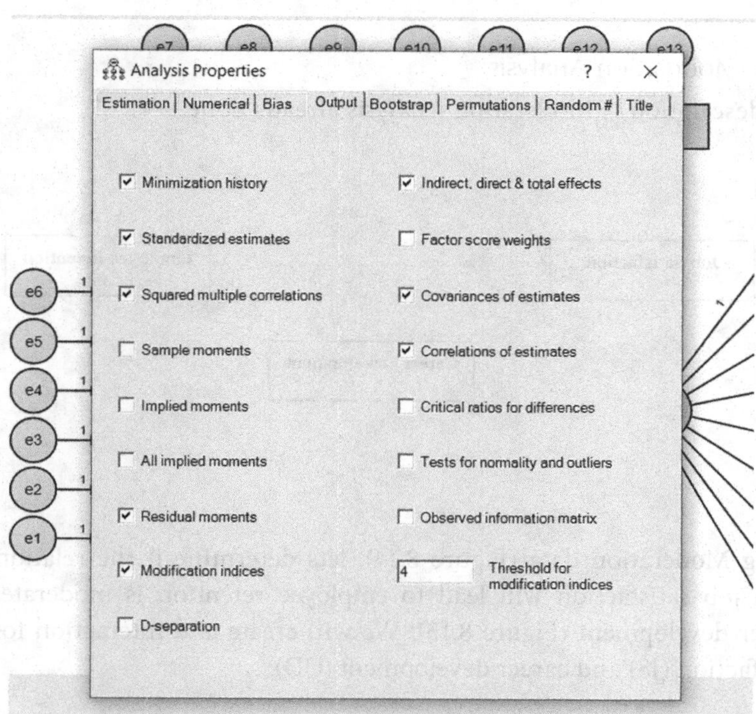

Then run the model by clicking Calculate the Estimates.

Click File > Save and choose a folder for Amos Analysis. Then Amos will give the CFA results. To get a display of the output model: click View text and you will have a display for a number of outputs from SEM (Figure 8.12). Within the Estimates, go to Matrices, and you will have Total Effects and Indirect Effects.

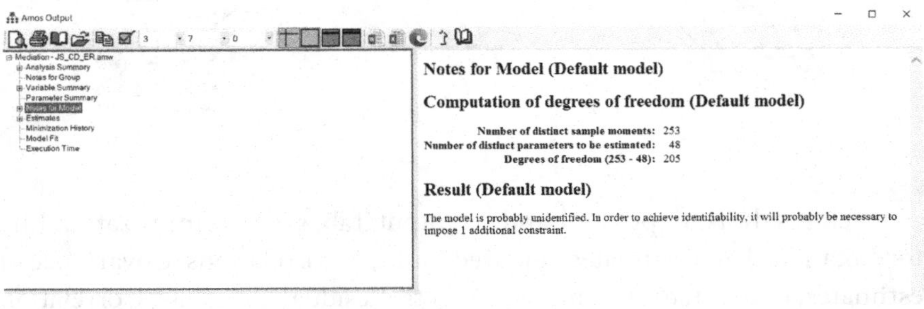

8.7.6 Moderation Analysis

The description of Moderation Analysis already done.

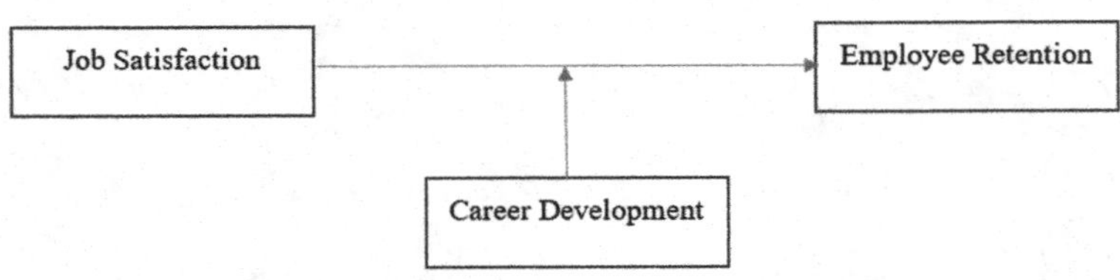

Using Moderation data (Figure 8.14), lets determine if the relationship from job satisfaction will lead to employee retention is moderated by career development (Figure 8.13). We will create and interaction for job satisfaction (JS) and career development (CD).

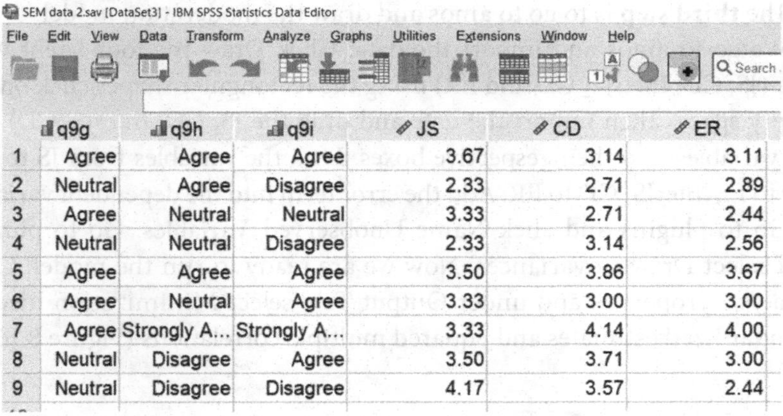

SEM data 2.sav [DataSet3] - IBM SPSS Statistics Data Editor

File Edit View Data Transform Analyze Graphs Utilities Extensions Window Help

Search

	q9g	q9h	q9i	JS	CD	ER
1	Agree	Agree	Agree	3.67	3.14	3.11
2	Neutral	Agree	Disagree	2.33	2.71	2.89
3	Agree	Neutral	Neutral	3.33	2.71	2.44
4	Neutral	Neutral	Disagree	2.33	3.14	2.56
5	Agree	Agree	Agree	3.50	3.86	3.67
6	Agree	Neutral	Agree	3.33	3.00	3.00
7	Agree	Strongly A...	Strongly A...	3.33	4.14	4.00
8	Neutral	Disagree	Agree	3.50	3.71	3.00
9	Neutral	Disagree	Disagree	4.17	3.57	2.44

The first step is **mean centering of the variables** in the data to help minimize high collinearity and makes interpretation of results easier. We will need to mean center the independent variable and the moderator variable before creating the product (interaction) term. We go to SPSS and get the mean for independent and moderator variable. We go to: Analyze > Descriptive Statistics > Frequencies and select the two variables then estimate the mean.

We then create new variables for JS and CD where we subtract the mean for JS and CD, respectively. We go to: Transform > Compute variable > put the expression for subtracting mean from each value JS and CD then create the interaction term for centered values for JS and CD.

The second step is to create the product terms of the centered variables. We will now have data set for moderation analysis (Figure 8.15).

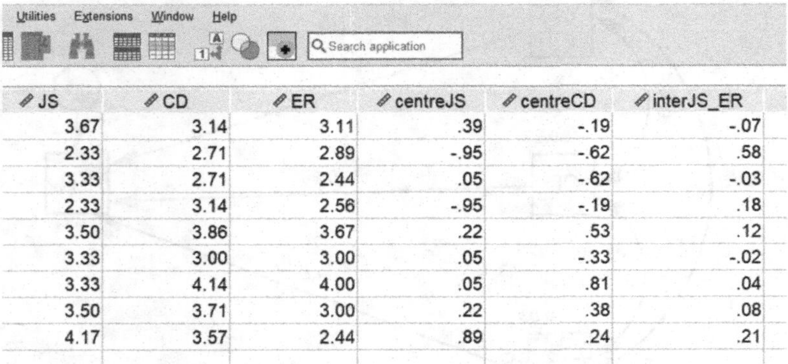

Utilities Extensions Window Help

Search application

JS	CD	ER	centreJS	centreCD	interJS_ER
3.67	3.14	3.11	.39	-.19	-.07
2.33	2.71	2.89	-.95	-.62	.58
3.33	2.71	2.44	.05	-.62	-.03
2.33	3.14	2.56	-.95	-.19	.18
3.50	3.86	3.67	.22	.53	.12
3.33	3.00	3.00	.05	-.33	-.02
3.33	4.14	4.00	.05	.81	.04
3.50	3.71	3.00	.22	.38	.08
4.17	3.57	2.44	.89	.24	.21

The **third step** is to go to amos and draw the moderation model.

We go to amos and import the data. Click Draw the four latent variable (JS, CD, InterJS_CD and ER) using the rectangular shapes click on the empty space. Then import the date and drag the JS, CD, InterJS_CD) and ER variables into their respective boxes. Link the variables from JS to ER, CD to Er, interJS_CD to ER. Add the error term into the dependent variable.

Go to plugins and click Name Unobserved Variables. Go to plugins and select Draw Covariances. Now we are ready to run the model. Click Analysis properties and under Output Tab, select Minimization history, Standardized estimates and Squared multiple correlations (Figure 8.16).

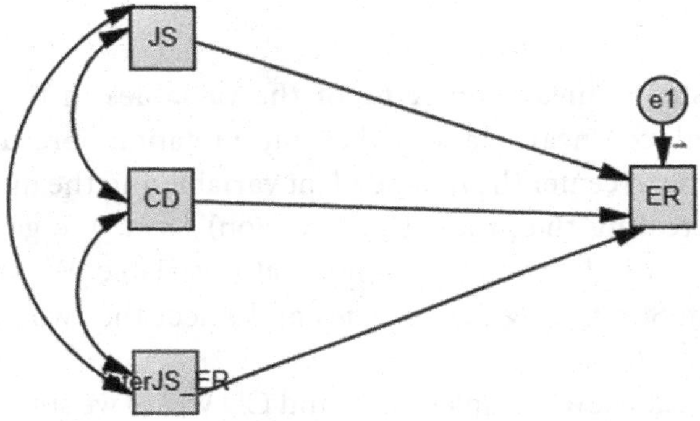

Click Calculate estimates and save the file. Amos will give the moderation results (Figure 8.17).

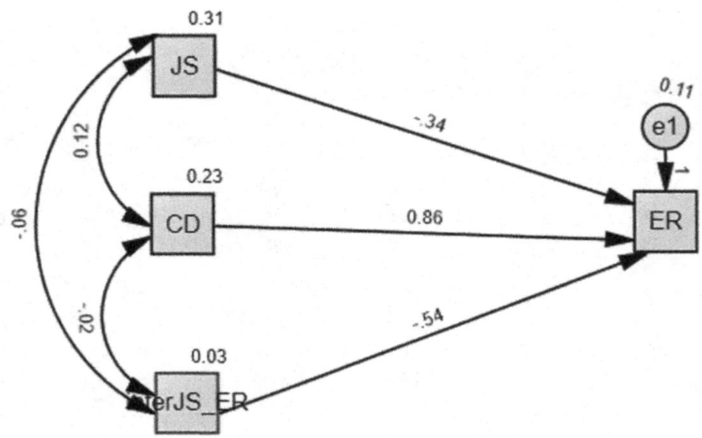

Click View texts to view your outputs (Figure 8.18).

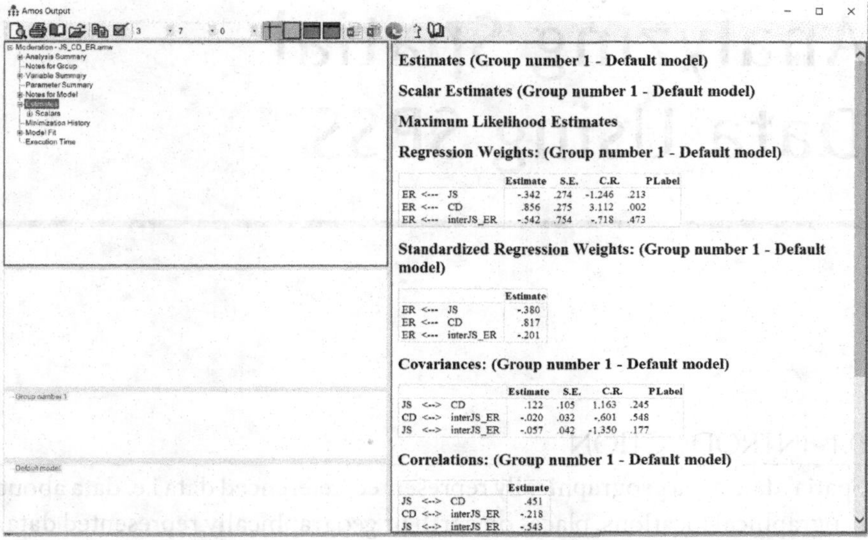

Go to Estimates and you will see the results indicating there is no moderation effect since the interaction term is not statistically significant ($p = 0.177$). The moderation effect is also negative.

8.8 PRACTICE EXERCISE

1. Classify the structural equation modeling techniques for analyzing data.
2. Using relevant SPSS data, perform and interpret the following analysis:

   a. CFA.

   b. Path analysis (mediation, moderation).

CHAPTER 9

Analyzing Spatial Data Using SPSS

9.1 INTRODUCTION

Spatial data are a geographically represented/referenced data i.e. data about geographical locations, places on earth or geographically represented data. It's also known as geospatial data.

Spatial statistics also involves techniques for analyzing spatial data.

Spatial statistics as a branch of statistics is used to analyze data that are observed on a multidimensional surface.

There are three main areas of spatial statistics: geostatistical data, lattice (areal) data and point patterns.

Spatial phenomena can either be viewed as *discrete objects* with clear boundaries or as a *continuous phenomenon* that can be observed everywhere, but that do not have natural boundaries. Examples of discrete spatial objects include a river, road, country, town or a research site. Examples of continuous phenomena, or "spatial fields", include elevation, temperature and air quality.

Spatial objects are usually represented by *vector data*. Such data consist of a description of the "geometry" or "shape" of the objects and normally also includes additional variables. Continuous spatial data (fields) are usually represented with a *raster data structure*.

9.1.1 Vector Data

The main vector data types are *points, lines and polygons*. In all cases, the geometry of these data structures consists of sets of coordinate pairs (x, y).

DOI: 10.1201/9781003386636-9

Points are the simplest case. Each point has one coordinate pair and *n* associated variables.

Line refers to a set of one or more polylines (connected series of line segments). Lines are represented as ordered sets of coordinates (nodes).

A *network* (e.g. a road or river network), or spatial graph, is a special type of lines geometry where there is additional information about things like flow, connectivity, direction and distance.

A *polygon* refers to a set of closed polylines. The geometry is very similar to that of lines, but to close a polygon, the last coordinate pair coincides with the first pair.

Examples of vector file formats include Shapefiles.

9.1.2 Raster Data

Raster data are commonly used to represent spatially continuous phenomena such as elevation. A raster divides the world into a grid of equally sized rectangles (referred to as cells or, in the context of satellite remote sensing, pixels) that all have one or more values (or missing values) for the variables of interest.

Continuous surface data are sometimes stored as triangulated irregular networks (TINs).

Examples of raster file formats include JPEG and PNG, among others.

9.1.3 Coordinate Reference System

A very important aspect of spatial data is the coordinate reference system (CRS) that is used.

9.1.3.1 Angular Coordinates

The earth has an irregular spheroid-like shape. The natural coordinate reference system for geographic data is longitude/latitude. This is an angular coordinate reference system. The latitude φ (phi) of a point is the angle between the equatorial plane and the line that passes through a point and the center of the Earth. Longitude λ (lambda) is the angle from a reference meridian (lines of constant longitude) to a meridian that passes through the point.

Before one can perform various spatial analyses and visualizations, we must convert the data into *spatial objects*. This is what is referred to as spatial data manipulation.

The main spatial data analysis techniques include point pattern analysis (PPA), spatial autocorrelation analysis and spatial regression analysis.

9.2 VISUALIZING SPATIAL DATA

Before diving into formal analysis, visualize your spatial data to get a sense of the distribution and patterns. Spatial data are visualized using maps.

We can generate geographical maps in SPSS using the path: Graphs > Graphboard Template Chooser > under Detailed Tab select Choropleth Means.

Creating maps with the Graphboard Template Chooser in SPSS involves using the Chart Builder to select a map template, customizing it and then displaying spatial data visually.

9.3 SPATIAL POINT PATTERN ANALYSIS

Point pattern analysis (PPA) focuses on the analysis, modeling, visualization and interpretation of point data. Methods for point pattern analysis include *Descriptive Statistics, Distance-Based Measures* and *Density-based Measures.*

9.3.1 Descriptive Statistics

In PPA, descriptive statistics provide a summary of the basic characteristics of a point pattern, such as its central tendency and dispersion.

A very basic form of point pattern analysis involves summary statistics such as the mean center, standard distance and standard deviational ellipse.

The *mean center* is the average of x and y coordinate values. The mean center coordinates $(\overline{x}, \overline{y})$. It is computed as

$$\overline{x} = \frac{1}{n}\sum_{i=1}^{n} x_i .$$

$$\overline{y} = \frac{1}{n}\sum_{i=1}^{n} y_i .$$

where:

n is the total number of points.

x_i and y_i are the x and y coordinates of the ith point respectively.

Standard distance is the measure of the variance between the average distances of the features of the mean center. Standard distances are defined similarly to standard deviations. This indicator measures how dispersed a group of points is around its mean center.

Given the mean center coordinates $(\overline{x}, \overline{y})$, the standard distance D is computed as

$$D = \sqrt{\frac{1}{n}\sum_{i=1}^{n}(x-\overline{x})^2 + (y-\overline{y})^2}.$$

The *standard deviational ellipse* is a separate standard distance for each axis. Standard deviational ellipses are used to calculate separate standard distances for two perpendicular axes. To compute the standard deviational ellipse, we first calculate the covariance matrix, then calculate the eigenvalues and eigen vectors of the covariance matrix. The lengths of the major and minor axes are then computed and then calculate the orientation of the ellipse and lastly the eccentricity of the ellipse.

9.3.2 Distance-Based Measures

Distance-based measures analyze the spatial distribution of points using distances between point pairs, and they are often considered a direct indicator of the second-order property.

9.3.2.1 Nearest-Neighbor Distance

Nearest-neighbor distance (NND) is the distance between a point and its closest neighboring point. NND is also known as the first-order nearest neighbor. In addition, distance can be calculated for the *k*th nearest neighbor, which is called the *k*th-order NN or KNN. The mean of NND between all point pairs is used as a global indicator to measure the overall pattern of a point set.

9.3.2.2 Distance Functions

G Function: The G function is the simplest one, which calculates the cumulative frequency distribution of the NND of a point pattern.

F Function: The F function first generates a few random points (denoted as P) in the study area, and then it determines the minimum distance from each random point in P to any original points (denoted as O) in the study area.

K Function: The G and F functions only consider the nearest neighbor for each point and ignore the distances to other points, so they cannot be used to analyze point patterns at multiple scales (distances). The Ripley's K function is a powerful approach to identify the multiscale patterns of points. The three aspects/steps/factors of Ripley's K function are (i) Construct a circle with a radius d around each point i; (ii) count the total number (n) of points that fall inside any of the circles (excluding the points at the circle centers) and (iii) increment d by a small fixed amount and repeat the first two steps.

9.3.3 Density-based Measures

Density measures can be divided into two categories: global density and local density. Global density refers to the ratio between the observed number of points relative to the size of the study area. Local density, however, shows varying point densities at different locations in the study area. The two most commonly used density-based measures are quadrat density and kernel density.

9.3.3.1 Quadrat Density

For quadrat density analysis, the study area is divided into smaller subregions (i.e. quadrats), and then the point density is computed for each subregion. Quadrant density for each quadrant can generally be computed as

$$QD_q = \frac{n_p}{A_q}.$$

where:

A_q is the area of the quadrant q computed as $A_q = \frac{(x_{max} - x_{min})(y_{max} - y_{min})}{k}$.
n_q is the number of points within quadrant q.
k is the number of quadrants that the study area has been divided into.

9.3.3.2 Kernel Density

Unlike quadrat density analysis, which assumes that the density of events is uniform within each quadrat, kernel density estimation (KDE) is based on the assumption that every location has a density, and the estimate of densities not only relies on the occurrence of events, but also a predefined mathematical equation (i.e. the kernel). More specifically, it estimates the local density of points in a nonparametric and continuous way by counting the number of events in a region (i.e. the search window) that is centered at the location where the density is calculated.

9.4 SPATIAL AUTOCORRELATION

Spatial autocorrelation analysis assesses spatial autocorrelation to identify if there are any spatial patterns in the data. Spatial autocorrelation measures the degree of similarity between the values of a variable at different locations.

There are several measures of spatial autocorrelation: Geary's index; Moran's index; Interval attributes; point, line and raster objects; Ordinal attributes; Nominal attributes; Variograms and correlograms. One popular test of spatial autocorrelation is the Moran's I test.

Moran's I values ranges between –1 and 1. –1 value indicates perfect dispersion (perfect negative spatial autocorrelation), 0 indicates spatial randomness (no spatial autocorrelation) and 1 indicates perfect clustering (perfect positive autocorrelation).

The Moran's I is computed as

$$I = \frac{n}{W}\frac{\sum_{i=1}^{n}\sum_{j=1}^{n} w_{ij}(x_i - \bar{x})(x_j - \bar{x})}{\sum_{i=1}^{n}(x_i - \bar{x})^2}.$$

where:

n is the number of observations at locations $i = 1, 2, \ldots, n$.
$\bar{x}$ is the mean value of the observations.
w_{ij} is the spatial weight matrix.
W is the sum of all elements in the spatial weight matrix.

We can use SPSS Syntax to compute the Moran's I index or use other statistical software such as R and Python.

9.5 SPATIAL INTERPOLATION

Spatial interpolation is the activity of estimating values of spatially continuous variables (fields) for spatial locations where they have not been observed, based on observations. The statistical methodology for spatial interpolation, called *geostatistics*, is concerned with the modeling, prediction and simulation of spatially continuous phenomena. Interpolation can be categorized into two categories: *deterministic* and *statistical* interpolation methods.

9.5.1 Deterministic Approach to Interpolation

We will explore two deterministic methods: proximity (aka Thiessen) techniques and inverse distance weighted techniques (IDW).

Proximity interpolation: The goal is simple: Assign to all unsampled locations the value of the closest sampled location. This generates a tessellated surface whereby lines that split the midpoint between each sampled location are connected thus enclosing an area. Each area ends up enclosing a sample point whose value it inherits.

Inverse distance weighted (IDW): The IDW technique computes an average value for unsampled locations using values from nearby weighted locations. The weights are proportional to the proximity of the sampled points to the unsampled location and can be specified by the IDW power coefficient. The larger the power coefficient, the stronger the weight of nearby points that are estimated using the *z* value at an unsampled location *j*.

The IDW interpolation estimate from the *n* sample points with known z_i at locations (x_i, y_i) is computed as

$$z = \frac{\sum_{i=1}^{n} w_i z_i}{\sum_{i=1}^{n} w_i}.$$

where w_i is the weight assigned to each sample point *i*, computed as $w_i = \frac{1}{d_i^p}$, where d_i is the distance between the sample point *i* and the target location, and *p* is the user-defined power parameter that controls the influence of the distance on the interpolation and it usually ranges between 1 and 3.

9.5.2 Statistical Approach to Interpolation

The statistical interpolation methods include surface trend and kriging.

Trend surfaces: It may help to think of trend surface modeling as a regression on spatial coordinates where the coefficients apply to those coordinate values and (for more complicated surface trends) to the interplay of the coordinate values. We will explore a 0th order, 1st order and 2nd order surface trend.

The 0th order model is a level (horizontal) surface whose cell values are all equal.

The first order surface polynomial is a slanted flat plane.

The second order surface polynomial (aka quadratic polynomial) is a parabolic surface.

Ordinary kriging: This form of kriging usually involves four steps:

Removing any spatial trend in the data (if present), computing the experimental variogram (which is a measure of spatial autocorrelation), defining an experimental variogram model that best characterizes the spatial autocorrelation in the data, interpolating the surface using the experimental variogram and adding the kriged interpolated surface to the trend interpolated surface to produce the final output.

9.6 SPATIAL REGRESSION

Spatial regression is used to explore spatial associations between disease occurrences and other variables using spatial regression models like spatial lag models (SLM) or spatial error Models (SEM).

Spatial lag model is used to account for spatial dependencies or autocorrelation in spatial data. It is also known as spatial autoregressive models. The general expression for a spatial lag model is given by

$$Y = \rho WY + X\beta + \varepsilon.$$

where:

Y is the vector of n observations of the dependent variable.
X is a matrix of size $n \times k$ representing k independent variables.
β is a vector of coefficients for the independent variables.
ρ is the spatial autoregressive parameter.
W is the spatial weights matrix.
ε is the vector of error terms.

Spatial error model accounts for spatial dependencies by modeling the spatial autocorrelation through an error term. The general expression for a spatial error model is given by

$$Y = \lambda WY + X\beta + \varepsilon.$$

where:

Y is the vector of n observations of the dependent variable.
X is a matrix of size $n \times k$ representing k independent variables.
β is a vector of coefficients for the independent variables.
λ is the spatial autoregressive parameter.
W is the spatial weights matrix.
ε is the vector of error terms.

Spatial regression models (like CAR models) to account for spatial autocorrelation in the disease data when analyzing relationships with other covariates.

We can generate spatial lag models and spatial error models for spatial data using the SPSS Syntax or use other statistical software like R and Python with special package and libraries for performing the analysis.

9.7 PRACTICE EXERCISE

1. Classify the spatial techniques for analyzing spatial data.
2. Using relevant SPSS data, perform and interpret the following analysis:
   a. Visualize spatial data.
   b. Point pattern analysis.
   c. Spatial autocorrelation.
   d. Spatial interpolation.
   e. Spatial regression.

Index

Note : The page preferences in **bold** and *italics* represents tables and figures respectively.

Printed in the United States
by Baker & Taylor Publisher Services